Theo Fett, Karl Günter Schell, Ethel C. Bucharsky, Gabriele Rizzi,
Pascal Hettich, Susanne Wagner, Michael J. Hoffmann

Consequences of hydroxyl generation by the silica/water reaction

Part I: Diffusion and Swelling

Schriftenreihe
des Instituts für Angewandte Materialien
Band 101

Karlsruher Institut für Technologie (KIT)
Institut für Angewandte Materialien (IAM)

Eine Übersicht aller bisher in dieser Schriftenreihe erschienenen Bände finden Sie am Ende des Buches.

Consequences of hydroxyl generation by the silica/water reaction

Part I: Diffusion and Swelling

by
Theo Fett
Karl Günter Schell
Ethel C. Bucharsky
Gabriele Rizzi
Pascal Hettich
Susanne Wagner
Michael J. Hoffmann

KIT Scientific Publishing

Impressum

KIT Scientific Publishing

Karlsruher Institut für Technologie (KIT)
KIT Scientific Publishing
Straße am Forum 2
D-76131 Karlsruhe

KIT Scientific Publishing is a registered trademark
of Karlsruhe Institute of Technology.
Reprint using the book cover is not allowed.

www.ksp.kit.edu

Print on Demand 2022 – Gedruckt auf FSC-zertifiziertem Papier

ISSN 2192-9963
ISBN 978-3-7315-1148-9
DOI 10.5445/KSP/1000136481

In Memory of

Sheldon M. Wiederhorn

The authors are very sad about the death of their old friend and colleague Shelley Wiederhorn. It was he who drew our interest to the interesting scientific aspects of the influence of water on silica glasses and its mechanical and fracture mechanical consequences. The results of the subsequent close collaboration have been presented in a number of joint scientific publications in scientific journals and oral presentations. Shelley visited us in our institute several times and always gave lectures on scientific items. We also met him often at national and international conferences.
Shelley retired from scientific work at the age of nearly 88. This booklet covers our common work, which reflects only a very small part of his wide-ranging activities. It is dedicated to him.
We will miss our friend Shelley very much. He will always be kept in honorable memory.

The staff of the
Institute for Applied Materials -
Ceramic Materials and Technology
Karlsruhe Institute of Technology (KIT)

DIFFUSION AND SWELLING IN SILICA

APPENDIX

Preliminary Remarks

The content of this compilation comprises our early work on water diffusion and swelling in silica between 2011 and 2018. Newer results will be contributed in a second booklet.

Water diffusing into silica surfaces gives rise for several effects on diffusion behaviour and mechanical properties. The most interesting observations from our point of view may be listed here.

(I) *In absence of externally applied stresses*:

a) Density reduction with increasing water concentration [1].

b) Increasing concentration of molecular water at silica surfaces [2].

c) Decrease of measured water diffusivity with time [3].

d) Change of the shape of the water concentration with time [3].

These points are covered in the present booklet (Part I).

(II) *In presence of externally applied loadings*:

a) Increasing diffusivity under externally applied tensile stresses and a decrease in compression [4].

b) Decrease of water solubility in tension and an increase in compression [4].

c) Increased water concentration in tension at high temperatures [5].

d) Strongly increasing strength at high tensile stresses even under very low water vapour pressure [6].

These issues are further elaborated in Part II.

For an explanation of all the findings listed under (I) there is only volume swelling by hydrogen generation an appropriate tool.

In order to give a transparent description, two main assumptions will be made:

(1) For an interpretation of the increase of diffusivity and water concentration under tensile loading, a further effect may play a role. Under tensile stresses, the nano-pores of the ring structure are possibly widened and diffusion and solubility may be enhanced. Such behaviour was observed in different materials and is interpreted in terms of the so-called free volume theory. This may also affect water diffusivity, generally described by the Doolittle equation [7], and solubility (see e.g. [8]). Due to the lack of results for silica, free volume effects must be ignored in the following considerations.

(2) Water molecules are believed to diffuse through silica glass in much the same way as the noble gases. Hence, the diffusion of water molecules is also expected to show little dependence on applied stress [5]. Following this reasoning, we believe that the molecular water in the glass is located in holes of molecular size in the silica glass network where the water does not contribute to the volume of the glass. Consequently, molecular water cannot contribute stress effects. Thus, the concentration of molecular water in the glass depends only on the vapour pressure in the external environment, but not on the stress in the glass.

In the APPENDIX, analytical and Finite Element results (FE) on cracks and pores in glass are compiled.

---

1 Shelby, J.E., "Density of vitreous silica," J. Non-Cryst. **349** (2004), 331-336.

2 Oehler, A., Tomozawa, M., Water diffusion into silica glass at a low temperature under high water vapor pressure, J. Non-Cryst. Sol. **347** (2004) 211-219.

3 Davis, K.M., Tomozawa, M., Water diffusion into silica glass: structural changes in silica glass and their effect on water solubility and diffusivity, J. Non-Cryst. Sol. **185** (1995), 203-220.

4 M. Nogami, M. Tomozawa, Diffusion of water in high silica glasses at low temperature, Phys. and Chem. of Glasses **25**(1984), 82-85.

5 A. Agarwal, M. Tomozawa, W. A. Lanford, Effect of stress on water diffusion in silica glass at various temperatures, J. Non-Cryst. **167**(1994), 139-148.

6 P.J. Lezzi, Q.R. Xiao, M. Tomozawa, T.A. Blanchet, C.R. Kurkjian, Strength increase of silica glass fibers by surface stress relaxation: A new mechanical strengthening method, J. of Non-Crystalline Solids **379** (2013) 95–106.

7 A.K. Doolittle, J. Appl. Phys. **22**(1951), 1471.

8 S. Neumann, G. Marom, Free-volume dependent moisture diffusion under stress in composite materials, J. Mat. Sci. **21**(1986), 26-30.

1 Diffusion in the absence of external stresses

1.1 Chemical reaction

Water penetrated into silica reacts with the silica network according to

$$\equiv\text{Si-O-Si}\equiv +\text{H}_2\text{O} \leftrightarrow \equiv\text{SiOH}+\text{HOSi}\equiv \tag{1.1.1}$$

with the concentration of the hydroxyl $S = [\equiv\text{SiOH}]$ and that of the molecular water $C = [H_2O]$. The equilibrium constant of this reaction is at temperatures of $\theta<450°\text{C}$

$$k_1 = \frac{S}{C} \tag{1.1.2}$$

In the range of $90°\text{C}\leq\theta\leq300°\text{C}$ the equilibrium constant can be expressed by [1.1]

$$k_1 \cong 32.3 \exp\left(\frac{-10750 J/mol}{RT}\right) \tag{1.1.3}$$

where $T=\theta+273\text{K}$ is the absolute temperature, and R the gas constant. The equilibrium constant is at temperatures $\theta>500°\text{C}$

$$k_2 = \frac{S^2}{C} \tag{1.1.4}$$

1.2 Water diffusion

Due to diffusion, concentrations of both molecular water, C, and hydroxyl water, S, decrease with increasing distance, z, from the glass surface. If the reaction given by eq.(1.1.1) is in *equilibrium*, the diffusion process is governed by the diffusion differential equation:

$$\frac{\partial C}{\partial t} = D\frac{\partial^2 C}{\partial z^2} \tag{1.2.1}$$

where D is the effective diffusivity that takes into account the chemical reaction given by eq.(1.1.1) [1.2]. The diffusivity shows an Arrhenius type of temperature dependence. Solution of the diffusion equation requires an appropriate boundary condition, very often chosen as constant surface concentration of molecular water:

$$C(z=0,t) \equiv C(0) = C_0 = \text{constant} \tag{1.2.2}$$

For a semi-infinite body, this assumption results in the following equation as a solution of the diffusion equation, *i.e.*, eq.(1.2.1):

$$C = C_0 \operatorname{erfc}\left(\frac{z}{2\sqrt{Dt}}\right) \tag{1.2.3}$$

which is well established for liquid water as the environment [1.3].

In the case of water vapour as the environment, the behavior of water at the silica interface does not follow eq.(1.2.2). Experimental results show that the concentration of water is not a constant at the surface, but increases continuously with time, until it reaches the solubility limit of water in the silica glass [1.4, 1.5].

Figure 1.1a shows both the molecular water and hydroxyl water concentrations of silica glass at the surface ($z = 0$) for a temperature of 250°C and a vapour pressure of 39 atm [1.6]. The fact that the equilibrium ratio between ≡SiOH and H_2O is achieved in these experiments is supported by the fact that the ratio, S/C, slightly below 3 is reached after a relatively short exposure time, ~200 h, Fig. 1.1b. However, the fact that $C(0)$ has not achieved a constant value in Fig. 1.1a suggests that the solubility limit of water in the glass has not been realized.

The experimental results in Fig. 1.1a [1.6] for $C(z=0, t)$ are a clear contradiction of the assumption of a constant surface value for the water concentration. One cannot assume that the surface concentration of water, $S(z=0)$ or $C(z=0)$, stays constant under the conditions used for the experiment. There is, however, an incredible agreement with the thermal analogue of thermal shock behaviour under heat transfer boundary conditions, governed by a heat transfer coefficient. This calls for a surface condition

$$\frac{dC}{dz} = \frac{h}{D}(C - C_0) \text{ at } z=0, \tag{1.2.4}$$

where again C_0 is the concentration of molecular water reached at z=0 for $t\rightarrow\infty$. Following the suggestion by Doremus (Section 4.7 in [1.7]), the phenomenological parameter h in (1.2.4) may be interpreted as a *reaction parameter* for a slow surface reaction that limits the entrance of molecular water species.

On the other hand, a simpler phenomenological description is possible by assuming that a barrier exists to the transport of water across the surface of the glass (for a physical description of such a barrier see Section 1.3). The barrier gives rise to a *mass transfer coefficient* for diffusion, which slows the passage of water into the glass. Each of the assumptions yields the same set of mathematical equations.

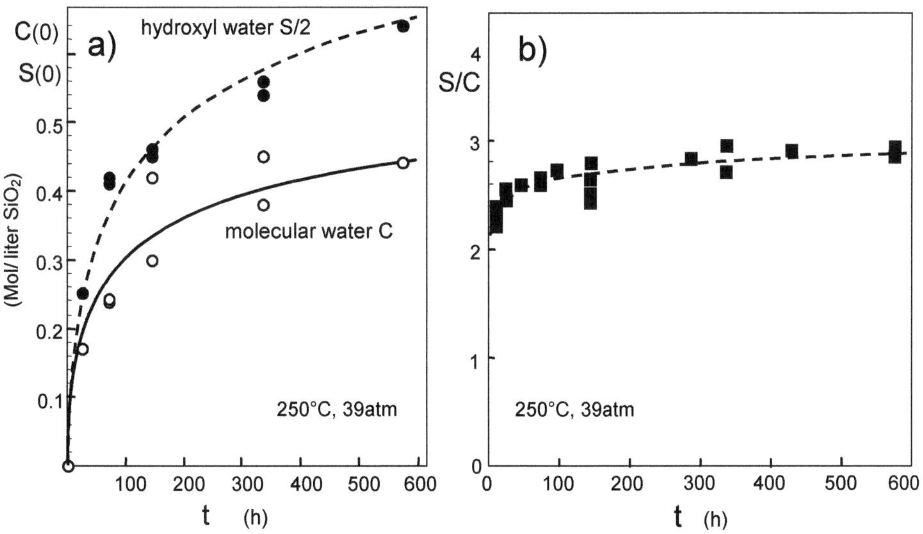

Fig. 1.1 a) Molecular water (C) and hydroxyl water content ($S/2$) at the surface of silica for 250°C at an equilibrium vapour pressure of 39atm, b) ratio S/C from [1.6].

As shown by Carslaw and Jaeger (Section 2.7 in [1.8]), the concentration profile, $C(z)$ resulting from the boundary condition for a semi-infinite body is given by

$$C(z,\, t)/C_0 = \mathrm{erfc}\left[\frac{z}{2\sqrt{Dt}}\right] - \exp\left[\frac{h}{D}z + \frac{h^2}{D}t\right]\mathrm{erfc}\left[\frac{z}{2\sqrt{Dt}} + h\sqrt{\frac{t}{D}}\right] \tag{1.2.5}$$

At the surface, z=0:

$$C(0,\, t)/C_0 = 1 - \exp\left[\frac{h^2}{D}t\right]\mathrm{erfc}\left[h\sqrt{\frac{t}{D}}\right], \tag{1.2.6}$$

For reasons of simplicity, we introduce a normalized dimensionless time τ and normalized depth coordinate ζ, defined by

$$\tau = \frac{h^2}{D}t\,;\ \zeta = \frac{z}{\sqrt{Dt}} \tag{*}$$

Equations (1.2.5) and (1.2.6) then read

$$\frac{C(\zeta,\tau)}{C_0} = \mathrm{erfc}\left[\frac{\zeta}{2}\right] - \exp[\zeta\sqrt{\tau} + \tau]\mathrm{erfc}\left[\frac{\zeta}{2} + \sqrt{\tau}\right] \tag{1.2.5a}$$

and

$$C(0,\, \tau)/C_0 = 1 - \exp[\tau]\mathrm{erfc}[\sqrt{\tau}], \tag{1.2.6a}$$

The C-measurements of Fig. 1.1a, carried out on specimens of the order of 1mm thickness, were fitted for the molecular water according to eq.(1.2.6). The result is

C_0 = 0.62 [0.523, 0.72] (mol H_2O/litre SiO_2), $h/\sqrt{D}$ = 0.075/√h [0.046, 0.104] (I)

with the 90% Confidence Intervals given in brackets. This result is introduced into Fig. 1.1a as the solid curve.

The measurements of Fig. 1.2 were carried out by Davis and Tomozawa [1.4] and Wakabayashi and Tomozawa [1.9] at a temperature of 350°C under 355 Torr water vapour pressure. The specimen widths, 2 W, ranged from 200 μm to 1000 μm. Since the maximum penetration of water was less than 70 μm (Fig. 4a in [1.4]), the solutions by eqs.(1.2.5) and (1.2.6) are applicable for these diffusion results.

At 350°C, the equilibrium of the S/C-ratio should be reached earlier than in the 250°C tests, because of the higher temperature. Consequently, S/C should be constant as suggested by the experiment carried out at 250°C and indicated in Fig. 1.2. The data points were obtained from IR-absorbance measurements and represent the IR-absorbance of the fundamental OH stretching band at wave number 3673 cm^{-1} per light path length through silica. In this case, the hydroxyl distribution and the molecular water distribution can be described by the same equation:

$$\frac{S(z,t)}{S_0} = \frac{C(z,t)}{C_0}, \tag{1.2.7}$$

where S_0 is the saturation value of the hydroxyl concentration for $t \to \infty$.

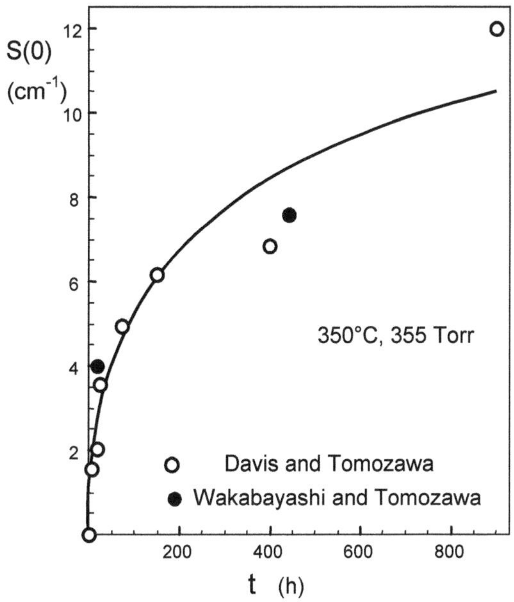

Fig. 1.2 Hydroxyl surface concentration after hydration at 350°C in water vapour at 355mm Hg; measurements were made by Tomozawa et al.[1.4, 1.9], (ordinate scaled in terms of the IR absorption coefficient).

By curve fitting eq. (1.2.6) to the data points in Fig. 1.2, the following parameters were determined:

$$S_0 = 18.9\,\text{cm}^{-1}\,[8.3, 29.5], \qquad h/\sqrt{D} = 0.0315/\sqrt{\text{h}}\,[0.0031, 0.060] \tag{II}$$

The result is introduced in Fig. 1.2 as the continuous curve.

For a temperature of 200°C, surface equilibrium between $C(0)$ and $S(0)$ is not expected in the short exposure times of 5-20h reported by Helmich and Rauch [1.5]. The hydroxyl content at the surface, $S(0)$, develops with time according to eq.(4.25) in reference [1.7]:

$$\frac{dS(0,\,t)}{dt} = \chi\, C(0,\,t) - \mu S(0,\,t) \tag{1.2.8}$$

Re-arranging yields a linear differential equation of first order

$$\frac{dS(0,\,t)}{dt} + \mu S(0,\,t) = \chi\, C(0,\,t) \tag{1.2.8a}$$

The two coefficients μ and χ can be related to the equilibrium constant k for the S/C-ratio, the ratio of saturation concentrations

$$\frac{S_0}{C_0} = k \tag{1.2.9}$$

From eq.(1.2.8) it follows for $t\to\infty$: $\mathrm{d}S/\mathrm{d}t=0$, and consequently

$$\chi = \mu\frac{S_0}{C_0} = \mu k \tag{1.2.10}$$

It should be noted that the equilibrium constant k is stress-dependent. It is well known from the work of Le Chatelier [1.10] that chemical reactions that exhibit a change in volume will be sensitive to the ambient pressure. This also holds for mechanical stresses as will be outlined in Sections 5 and 6. When k is increased for instance by mechanical stress, the ratio $k=\chi/\mu$ must increase, too. It has to be expected that the forward reaction rate χ will be increased, the reverse reaction rate, μ, will be reduced. Especially for tensile stresses in the GPa region, the stress effect on reaction rate can become strong.

Since the concentration $C(0,\,t)$ on the right-hand side of eq. (1.2.8a) is known from eq. (1.2.6) as the surface boundary condition, and the surface boundary condition $S(0,\,t)$ is known, too, namely $S(0,\,t) = k\cdot C(0,\,t)$. The solution (1.2.8a) can be obtained by using the Mathematica Routine *DSolve* [1.12] resulting in

$$\frac{S(0,t)}{C_0} = \frac{k}{\alpha^2+\mu}\left\{\alpha^2 + \mu(1-\exp[\alpha^2 t]\mathrm{erfc}[\alpha\sqrt{t}]) - \alpha\exp[-\mu t](\alpha + \sqrt{\mu}\,\mathrm{erfi}[\sqrt{\mu t}])\right\} \tag{1.2.11}$$

with the abbreviation $\alpha = h/\sqrt{D}$ and the *imaginary error function* erfi. The total water content in molar units C_w is then given by

$$C_w(0,t) = C(0,t) + \tfrac{1}{2}S(0,t)\ ,\quad C_{w,0} = C_0 + S_0/2 \tag{1.2.12}$$

In Fig. 1.3 the molecular, hydroxyl and total water concentrations at the surface are plotted versus normalized time. All concentrations are normalized to their saturation values.

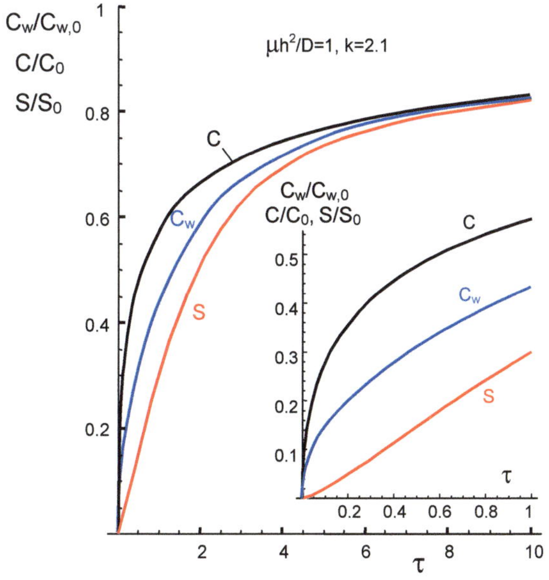

Fig. 1.3 Surface values of the water species *C*, *S* and the total water C_w as a function of time for an equilibrium constant of k=2.1, valid for 200°C.

Water concentration measurements were carried out on silica by Helmich and Rauch in [1.5]. The silica specimens were soaked in saturated water vapour at 200°C, 15.3 bar (1.53MPa). Fig. 1.4a shows the surface concentration of the water from H-measurements using the technique of Nuclear Reaction Analysis [1.5]. From data fitting and using S_0/C_0=2.1 it results:

$$C_0 = 1.71\times 10^{20}/\mathrm{cm}^3,\ \alpha = 1.902/\sqrt{\mathrm{h}},\ \mu = 0.130/\mathrm{h} \tag{III}$$

From C_0 we obtain: $S_0 = 2.1\times C_0 = 3.42\times 10^{20}/\mathrm{cm}^3$, $C_{w,0} = 2.05\ C_0 = 3.51\times 10^{20}/\mathrm{cm}^3$.

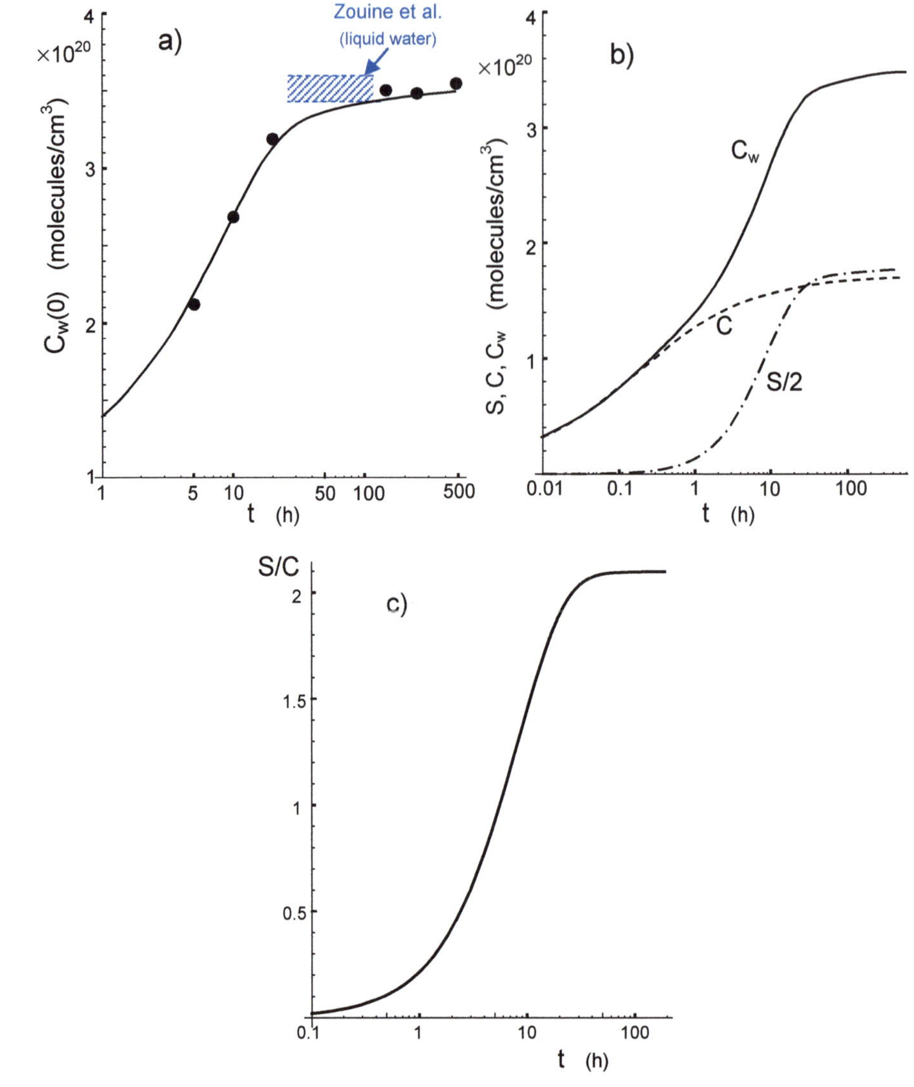

Fig. 1.4 a) Total water concentration C_w at the surface at a temperature of 200°C in saturated water vapour by Helmich and Rauch [1.5] (data points), Curve: eq.(1.2.12) fitted to the data, hatched area: Region of measurements by Zouine et al. [1.3] carried out in liquid water, b) hydroxyl and molecular water concentrations, c) S/C-ratio vs. time.

Since only six data points were available for the fitting analysis, and four parameters had to be determined, the results do not have high precision. Hence, confidence intervals were not determined. The two water species, C and $S/2$ are shown in Fig. 1.4b. Finally, the time-dependent S/C-ratio is represented in Fig. 1.4c. From this diagram, we can expect that S/C equilibrium is reached after about 30 h.

Surface concentration for tests in liquid water

Water concentrations in silica glass exposed to liquid water are reported by Zouine *et al.* [1.3] for temperatures ranging from 25°C to 200°C. For several of the temperatures, the authors measured water content after different times. The ratio of longer to shorter times ranged from about 2 to 4. The related concentrations ratios, C_{short}/C_{long}, are plotted in Fig. 1.5 as a function of temperature.

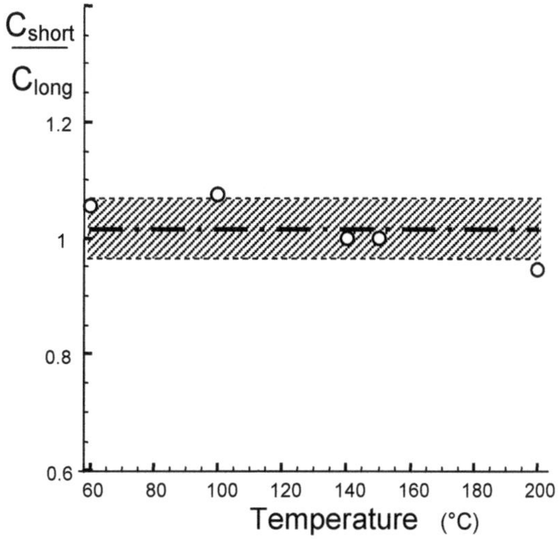

Fig. 1.5 Ratio of the surface water concentrations for "short" and "long" soaking times from Zouine et al. [1.3]; dash-dotted line: mean value, dashed lines: ±standard deviation.

The mean value was found to be $C_{short}/C_{long} = 1.015$ with the standard deviation SD $= 0.052$. The mean value is represented in Fig. 1.5 by the dash-dotted line, the region of ±SD by the hatched area. Thus, there is no indication for an increasing ratio of surface water concentration with increased soaking time as there is for water vapour. From this result we can conclude that the results from [1.3] represent saturation values. Consequently, a constant surface concentration of water and the use of eq.(1.2.3) is justified for the analysis.

It should be noted that not the high water pressure is responsible for the nearly time-independent water concentration for the tests in [3], since in the experiments by Oehler and Tomozawa [6], the saturation pressure was at 350°C clearly higher than in the tests by Zouine et al. [3]. Obviously the *direct contact* of silica and liquid water is responsible for a large value of $h/\sqrt{D}$.

1.3 Mass transfer coefficients

1.3.1 Parameter h/√D from literature data

First computations of the mass transfer parameter were reported in [1.11]. These results are based on total water uptake and surface water concentration. They may be addressed first.

The molecular water uptake per surface unit, m_C, after time t results by integrating the water profile of eq.(1.2.5) from z=0 to $z\to\infty$. With the help of Mathematica [1.12] we obtain

$$m_C = \int_0^\infty C(z)dz = C_0 \frac{D}{h}\left\{\frac{2}{\sqrt{\pi}}\frac{h}{\sqrt{D}}\sqrt{t} - \left(1-\exp\left(\frac{h^2}{D}t\right)\operatorname{erfc}\left(h\sqrt{\frac{t}{D}}\right)\right)\right\} \tag{1.3.1}$$

For large values of $h\sqrt{(t/D_0)}$ the water uptake tends asymptotically to

$$m_\infty \to C_0\left\{\frac{2}{\sqrt{\pi}}\sqrt{tD} - \frac{D}{h}\right\} \tag{1.3.2}$$

The time $\sqrt{t_0}$ at which the asymptote coincides with the $\sqrt{t}$ axis is

$$\sqrt{t_0} = \frac{\sqrt{\pi}}{2}\frac{\sqrt{D}}{h} \tag{1.3.3}$$

Under the assumption that equilibrium between molecular and hydroxyl water is reached even for short soaking times, eqs.(1.3.1) and (1.3.2) are valid also for the hydroxyl S. Uptake measurements by Wakabayashi and Tomozawa [1.9] for the hydroxyl water S are plotted in Fig. 1.6 by the solid and open circles.

In the first evaluation given in [1.11], we fitted eq.(1.3.1) to all the data and obtained the curves in Fig. 1.6a. The fitted parameters $h/\sqrt{D}$ are plotted in Fig. 1.7 as the open circles. An evaluation of the abscissa sections $\sqrt{t_0}$ in Fig. 1.6b via eq.(1.3.3) resulted in the triangles of Fig. 1.7.

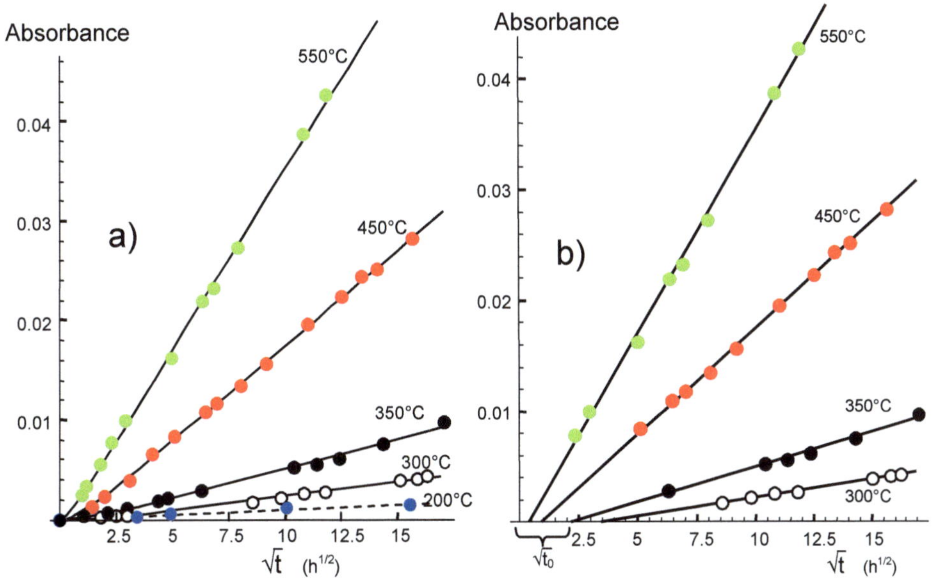

Fig. 1.6 a) Water uptake of silica by water vapour soaking at different temperatures and water vapour pressure of 355 Torr after Wakabayashi and Tomozawa [1.9], curves fitted to the data via eq.(1.3.1), b) determination of the asymptotes from results at large times using eq.(1.3.3).

The straight lines in this diagram yield

$$\frac{h}{\sqrt{D}} = A \exp\left(-\frac{Q}{RT}\right) \tag{1.3.4}$$

with the fitting parameters: A=527 (1/h$^{1/2}$), Q=38.1 (kJ/mol) for the circles and A=306 (1/h$^{1/2}$), Q=34.0 (kJ/mol) for the triangles.

The surface concentration values for molecular water obtained from the measurement by Oehler and Tomozawa [1.6] (solid circle) are also in good agreement with the uptake data. The surface hydroxyl measurements (S) by Wakabayashi and Tomozawa [1.9] result in a clearly lower value (square). The 90% CI are given by the bars.

All data in Fig. 1.7 apart from the square reflect directly or indirectly the behaviour of the molecular water. This holds of course for the solid circle since this value was determined directly from the surface history of molecular water.

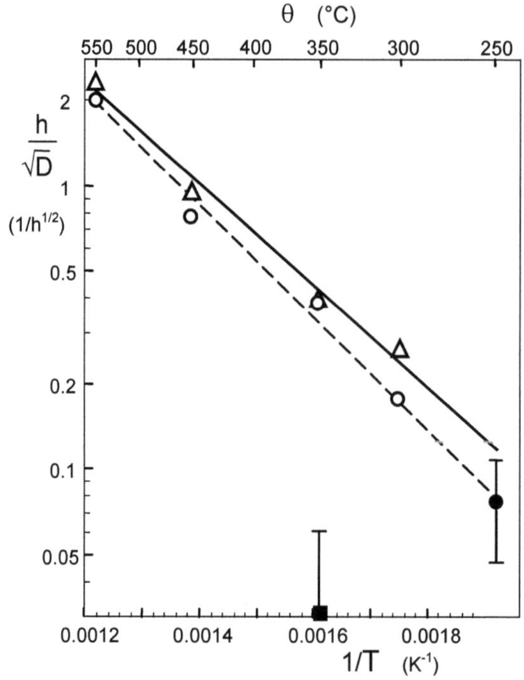

Fig. 1.7 Parameter $h/\sqrt{D}$ from [1.11] as a function of temperature for constant water vapour pressure of 355 Torr; circles: evaluation of the data in Fig. 1.6a by fitting to eq.(1.3.1), triangles: results from the asymptotic behaviour via eq.(1.3.3). Solid circle represents data from Fig. 1.1a (Oehler and Tomozawa [1.6], saturation pressure). Square gives result from Fig. 1.2 (Wakabayashi and Tomozawa [1.9], Davis and Tomozawa [1.4], 355 Torr).

On the other hand, the asymptotic evaluation of eqs.(1.3.1) and (1.3.3) is predominantly governed by the long-time data. The evaluation of Fig. 1.6a includes also some data measured at short times for which *S*/*C*-equilibrium might not have been reached. Since the majority of data points are very close to the asymptotes, the asymptotic behaviour prevails. Consequently, the *shape parameters* $h/\sqrt{D}$ for S (as measured in [1.9]) should be approximately the same as for C.

1.3.2 Re-evaluation and replenishment of data from literature

Considering again the data in Fig. 1.2a, large scatter in the measurements is noticeable and the deviations of the individual data points from the fitting curve are enormous. This holds especially for soaking times >400 h and is clearly caused by the high value at 900h, which appears extremely questionable.

Therefore, we excluded the 900h-value and fitted eq.(1.2.6) to the remaining data. The fit result was found as $h/\sqrt{D}$=0.11 [0.077,0.148] (1/√h). The related curve is introduced in Fig. 1.8a as the solid line. The dashed curve represents again the result from Fig. 1.2.
In the paper by Wakabayashi and Tomozawa [1.9] additional data of surface concentrations are reported. These results were obtained by stepwise removal of the surface via etching.

Figure 1.8b shows results for 550°C. These data were fitted with eq.(1.2.6) yielding $h/\sqrt{D}$ =0.66 [0.56, 0.75] (1/√h). For the other temperatures, only two data points were available so that the application of a fitting procedure was not possible. Table 1.1 compiles the new results that are also introduced in Fig. 1.9 by the red symbols, from [1.13].

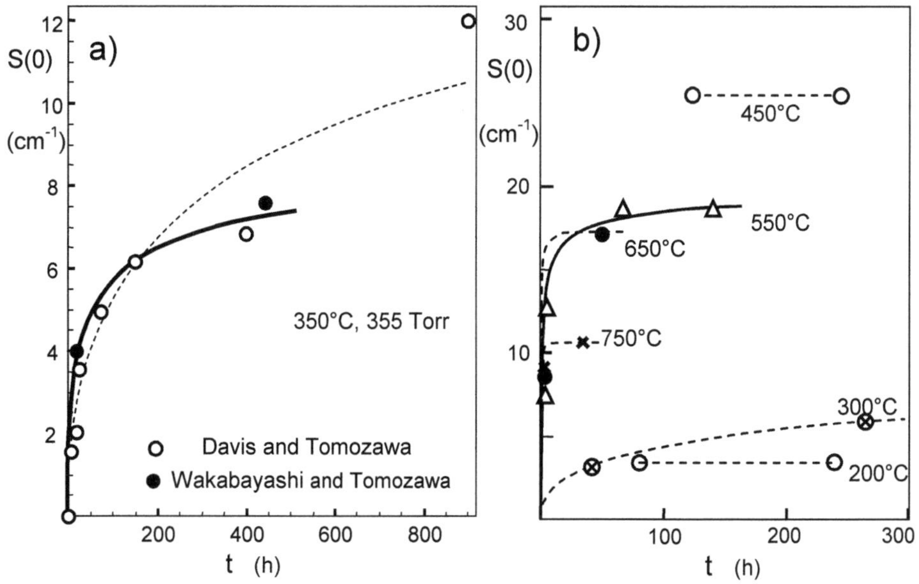

Fig. 1.8 a) Re-evaluation of strongly scattering data by Davis and Tomozawa [1.4] neglecting the data point at 900h, b) additional data by Wakabayashi and Tomozawa [1.9], evaluation from [1.13].

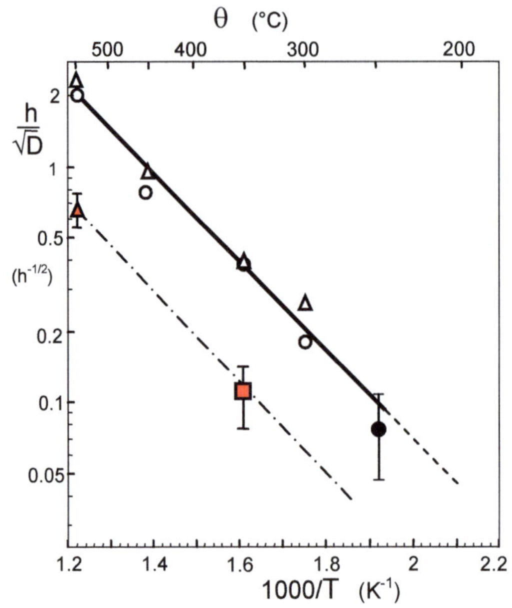

Fig. 1.9 Parameter $h/\sqrt{D}$ as a function of temperature from Fig. 1.7. Additionally plotted: parameters evaluated from Figs. 1.8a and 1.8b shown as red symbols.

The scatter of heat transfer parameters from literature may be caused by the different vapour pressures (saturation pressure and 355Torr), and procedures (global water uptake and stepwise surface removal).

In addition, the water-silica reaction is accompanied by the generation of *swelling strains* as will be outlined in detail in Section 2. Such swelling strains were observed in [1.1] for a GE-124 type-I glass by measuring the change of the curvature of silica disks after hot-water soaking.

Temperature (°C)	Basis data	$h/\sqrt{D}$ (1/√h)	Surface concentration
350	[1.4, 1.9]	0.11 [0.077, 0.148]	9.4 (cm^{-1})
550	[1.9]	0.66 [0.56, 0.75]	20.2 (cm^{-1})

Table 1.1 Additional mass transfer coefficients from curve fitting via eq.(1.2.6).

1.3.3 A possible reason for a limited water entrance

What we have demonstrated in the preceding Sections is that by assuming that silica glass surfaces present a barrier to the diffusion of water, we can duplicate the experimental diffusion results of Oehler and Tomozawa [1.6] and those of Helmich and Rauch [1.5]. We can also show that experimental diffusion results obtained in liquid water should be essentially different from those obtained in water vapour. The experiments of Zouine et al. [1.3] demonstrated the absence of a barrier when diffusion experiments are done in liquid water. These authors obtained the identical surface concentrations for specimens held for very different lengths of time at temperatures ranging from 23°C to 200°C. A barrier is observed only when silica is exposed to water vapour. Immersion in liquid water differs significantly from immersion in the vapour, because silica dissolves in the liquid [1.14, 1.15, 1.16], but not in the vapour [1.14]. Hence in liquid water, the surface is constantly being regenerated and any barrier that might form is dissolved away.

By comparison, specimens exposed to water vapour do not lose weight [1.14]. Not only that, their surfaces are still polished after exposure [1.6, 1.1] suggesting that whatever enters the surface during the exposure process, remains there when the specimen is removed from the moist environment.

Recent investigation of silica surfaces show that silica is more reactive chemically at the surface than in the bulk, which may account for the formation of a diffusion barrier at the surface. These surfaces do not dissolve away when exposed to water vapour. In contrast to the SiO_2-structure in the bulk of silica, under-coordinated Si and O atoms occur at the free surface because of missing neighboured atoms. These are more reactive so that when this surface is exposed to water vapour, water molecules react nearly instantaneously with the SiO_2 surface molecules to form SiOH as has been shown by molecular dynamic computations (see e.g. Mahadevan and Garofalini [1.17]).

The composition of a silica surface after contact with water molecules is illustrated in Fig. 1.10a. It is conceivable that such a layer can show a completely different diffusion behavior than the bulk material. Such a difference would be necessary to form a diffusion barrier.

When the silica surface is exposed to a water vapour environment, additional water molecules are docked at the hydroxyl groups. Water molecules are trapped at the "free" ends of the surface molecules for a short while resulting in a dynamic equilibrium: water molecules coming in from the environment are then captured by the surface, while water released from the surface is free to diffuse deeper into the bulk as has been studied by computer modelling of the diffusion process [1.18,1.19,1.20].

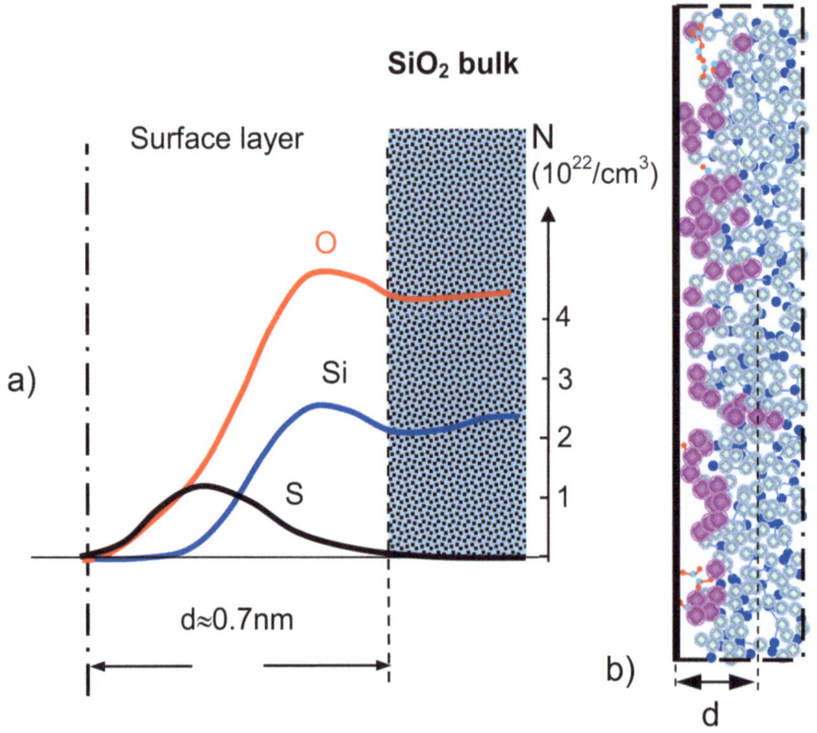

Fig. 1.10 a) At a free silica surface water reacts nearly instantaneously with under-coordinated silica species in a thin surface layer of about 0.7 nm thickness; per 2-coord O (and 3-coord Si) 1 H_2O molecule reacts to SiOH (curves according to Mahadevan and Garofalini [1.17]), b) surface layer of water molecules on a silica surface (pink: H of SiOH or H_2O, grey: O of silica; blue: Si).

A surface layer formed by this process, Fig. 1.10b, was developed by Mahadevan and Garofalini [1.17]. In Fig. 1.11a the hydroxyl concentration is given in weight percents (right ordinate). Unfortunately, the swelling behaviour of the under-coordinated Si and O atoms of silica is not known.

The measurements by Brückner, Shackelford and Shelby predominantly represent the behaviour of the bulk material with "correctly" coordinated oxygen and silicon. Therefore, we tentatively apply the dependency of volume swelling strain as a function of hydroxyl concentration, $\varepsilon_v(S)$, given in [1.21] as an approximation.

The hydrostatic swelling stresses can then be computed as shown in [1.21]. The result is represented by the left ordinate of Fig. 1.11a. Compressive hydrostatic stress reaches a maximum value of $\sigma_{sw,h} = -2.8\,\text{GPa}$.

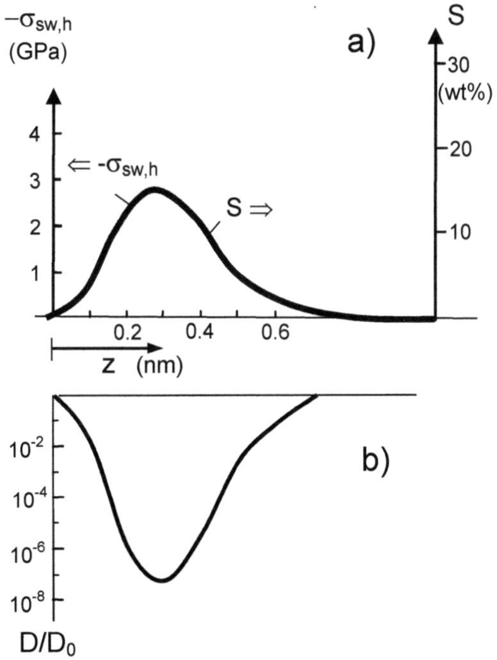

Fig. 1.11 Hydroxyl concentration (a) in weight percent, resulting in compressive hydrostatic swelling stresses (upper curve). The lower curve (b) shows the strongly reduced water diffusivity D, normalized on the diffusivity D_0 for stress-free silica at room temperature.

Due to such high compressive stresses, the water diffusivity must be strongly reduced within this thin layer. When swelling stresses are present, the diffusivity is given by the following equation:

$$D = D_0 \exp\left[\sigma_{sw,h} \frac{\Delta V_w}{RT}\right] \tag{1.3.5}$$

with the effective volume $\Delta V_w \cong 16$ cm³/mol as derived by Lezzi et al. [1.22].

Figure 1.12 schematically illustrates the consequences of the reduced diffusivity in the surface layer. In Fig. 1.12a the distribution of the diffusivity along the depth coordinate is plotted once more. Figure 1.12b shows the water profile for a fixed time. The water concentration directly at the surface, C_0, is assumed generally to be constant only depending on the partial pressure of water in the environment.

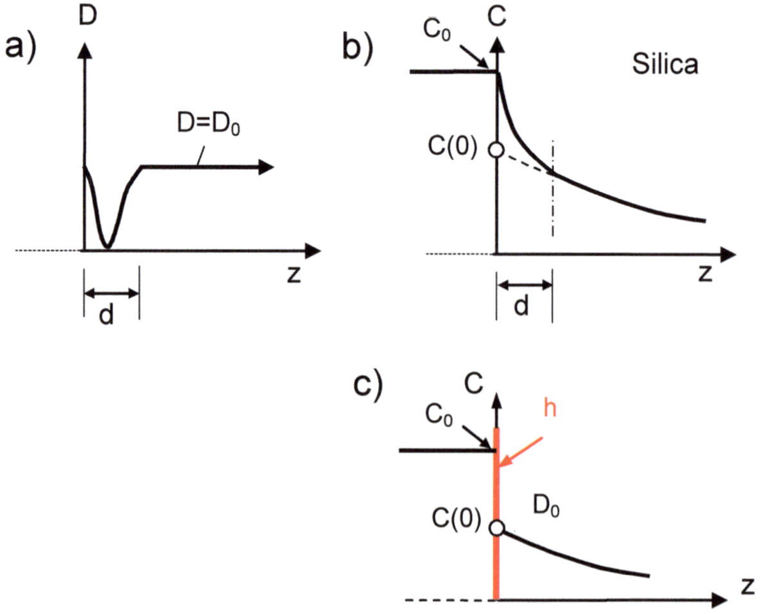

Fig. 1.12 a) Schematic representation of the diffusivity in the surface region, b) related water concentration profile, c) finite surface layer replaced by a surface with a finite mass transfer coefficient.

From experimental water concentration measurements as for instance performed by Oehler and Tomozawa [1.6], the distribution of C is obtained. Extrapolation of the measured curves to z=0 results in the dashed extension through the surface layer, Fig. 1.12b. The surface value of the extrapolated profile, $C(0)$, (indicated in Fig. 1.12b by the circle) must of course be smaller than C_0 and must increase with time as is found in the experiments.

The suppressed water transport within $0\leq z\leq d$ can be taken into consideration by replacing the thin layer by the surface ($d\rightarrow 0$) and introducing a finite mass transfer coefficient, h, into the surface boundary condition (Fig. 1.12c).

When silica is exposed to *liquid water*, dissolution of the glass into the water has to be expected. Dissolution of silica in water has been reported by a number of authors and is summarized in [1.23].

From these data we can conclude that in liquid water the silica surface in Fig. 1.10b and the highly stressed layer in Fig. 1.11 are continuously removed by dissolution. By contrast, in *water vapour* the transport of the SiOH from the surface is not possible. This is probably the reason for the difference in diffusion behaviour in liquid water and in water vapour environments.

1.4 Diffusion in thin specimens

1.4.1 Analytical solutions from literature

In the experimental and theoretical considerations in Section 1.2, diffusion in thick specimens was discussed. The diffusion zones were negligible compared to the thickness of the bulk material, $\sqrt{D\times t} << W$, so that the limit case of a half-infinite body is sufficiently fulfilled.

For *thin specimens* of width $2W$, the zone thickness may be comparable with the specimen thickness. Then the concentration distribution as a function of time reads according to [1.8]

$$C(z,t)=C_0\left(1-\sum_{n=1}^{\infty}\frac{2Wh'\cos[\alpha_n(z/W-1)]\sec(\alpha_n)}{Wh'(Wh'+1)+\alpha_n^2}\exp\left[-\alpha_n^2\frac{D}{W^2}t\right]\right)\;,\quad h'=h/D \qquad (1.4.1)$$

with the time-dependent surface concentration (z=0)

$$C_s(t) = C_0\left(1 - \sum_{n=1}^{\infty} \frac{2W\,h'}{Wh'(Wh'+1) + \alpha_n^2} \exp\left[-\alpha_n^2 \frac{D}{W^2} t\right]\right) \tag{1.4.2}$$

where C_0 is the saturation value and α_n, n=1, 2, ... are the positive roots of

$$\alpha \tan\alpha = W\,h' = Wh / D \tag{1.4.3}$$

In eqs.(1.4.1-1.4.3) the parameter Wh/D is equivalent to the Biot number B (sometimes also written Bi) for thermal problems:

$$Wh' = \frac{Wh}{D} \equiv B \tag{1.4.4}$$

Then the equations (1.4.1-1.4.3) read

$$C = C_0\left(1 - \sum_{n=1}^{\infty} \frac{2B\cos[\alpha_n(z/W - 1)]\sec(\alpha_n)}{B(B+1) + \alpha_n^2} \exp\left[-\alpha_n^2 \frac{\tau}{B^2}\right]\right) \tag{1.4.1a}$$

$$C_s(t) = C_0\left(1 - \sum_{n=1}^{\infty} \frac{2B}{B(B+1) + \alpha_n^2} \exp\left[-\alpha_n^2 \frac{\tau}{B^2}\right]\right) \tag{1.4.2a}$$

where now α_n, n= 1, 2, ... are the positive roots of

$$\alpha \tan\alpha = B \tag{1.4.3a}$$

The water concentration profiles as a function of time are shown for B=1 in Fig. 1.13a in normalized representation. After a normalized time of τ=0.03, the diffusion fronts from the opposite surfaces start to interfere noticeably.

The water concentration at the surface, $C(0)$, is given in Fig. 1.13b for several Biot numbers. At first sight, these results may appear somewhat astonishing since for increasing Biot numbers the surface concentrations at a fixed time τ decrease. For constant D and W, this says that C_0 decreases with increasing mass transfer coefficient h. In this context, it has to be noticed that the normalized time τ also includes the parameter h via $\tau \propto h^2$, so that for short normalized times, the curves become independent of h. This is visible from Fig. 1.13b.

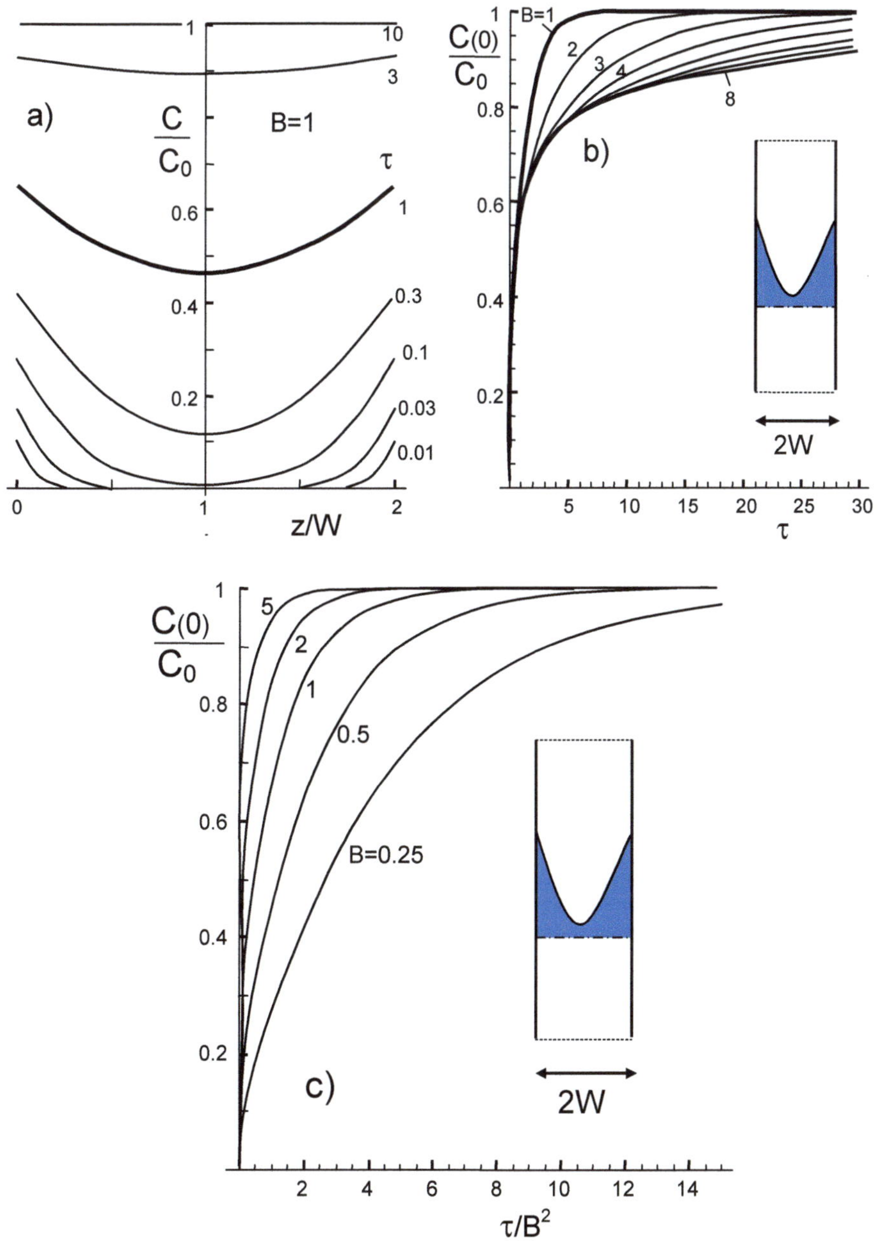

Fig. 1.13 a) Concentration profiles for $D \times t / W^2 = \tau / B^2 = 1$ and different normalized times τ, b) normalized surface concentrations C_s/C_0 vs normalized time for different Biot numbers, c) results with modified abscissa scaling τ/B^2 for small Biot numbers.

Figure 1.13c shows results of $C(0)$ for small values of B with the abscissa τ/B^2 instead of τ. These curves now show the properties to be expected intuitively.

Since the water uptake m_C is proportional to the average water concentration over the cross-section, it holds simply

$$\frac{m_C}{m_0} = \frac{\overline{C}}{C_0} \tag{1.4.5}$$

with the average water concentration $\overline{C}$:

$$\overline{C}(t) = \tfrac{1}{2W} \int_0^{2W} C(z,t)dz \tag{1.4.6}$$

and the saturation uptake m_0.

The molecular water uptake as a function of normalized time τ and Biot number is given in Fig. 1.14a and Fig. 1.14b for $\sqrt{\tau}$ and $\sqrt{\tau}/B$ chosen as the abscissa. As a characteristic feature of the effect of a limited water transfer coefficient *h,* the curves do not start linearly. This would only occur under the assumption that the surface concentration of molecular water would be constant during the water-soaking test, i.e. for $C(0,t)=C_0$=const.

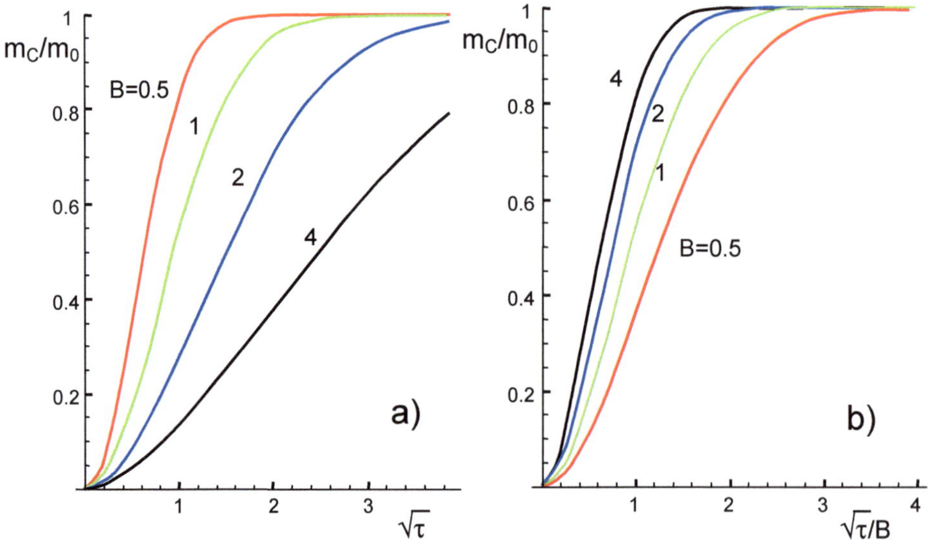

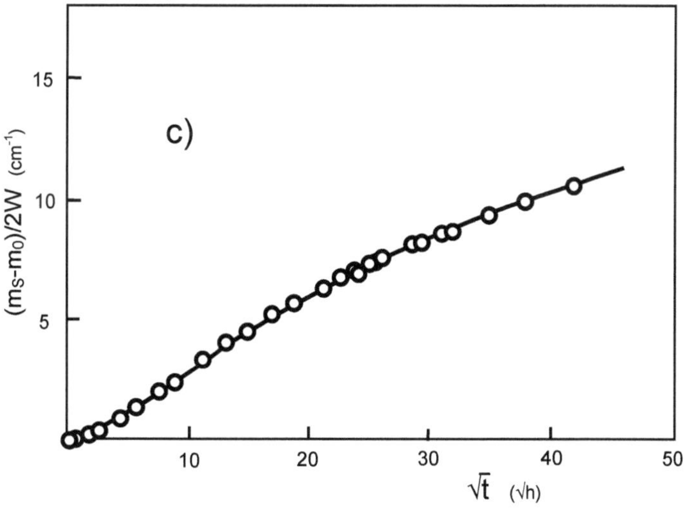

Fig. 1.14 a) Water uptake m_C/m_0 as a function of normalized time τ and Biot number B, b) the same results with different abscissa, c) measurements by Davis and Tomozawa [1.4].

In real tests (see Fig. 1.14c), the surface concentrations increase for short times $\propto\sqrt{t}$ and the layer thickness goes with $\sqrt{t}$, too. This makes that the uptake is proportional to time. In a plot with $\sqrt{t}$ as the abscissa, the uptake must start with a quadratic dependency. This effect is also visible from literature results by Davis and Tomozawa [1.4].

1.4.2 Results for high temperatures >500°C

The equilibrium constant of the reaction (1.1.1) is at high temperatures, T>500°C, represented by the ratio

$$k_2 = \frac{S^2}{C} \tag{1.4.7}$$

Results of S as a function of swelling stresses are given in Figs. 1.15a and 1.15b. The influence of Biot number is shown in Fig. 1.15c. In principle the same behaviour is visible as in the case of lower temperatures. In Figure 1.15 the asymptotically reached maximum hydroxyl content is denoted as S_0 given by

$$S_0 = \sqrt{k_2 C_0} \tag{1.4.8}$$

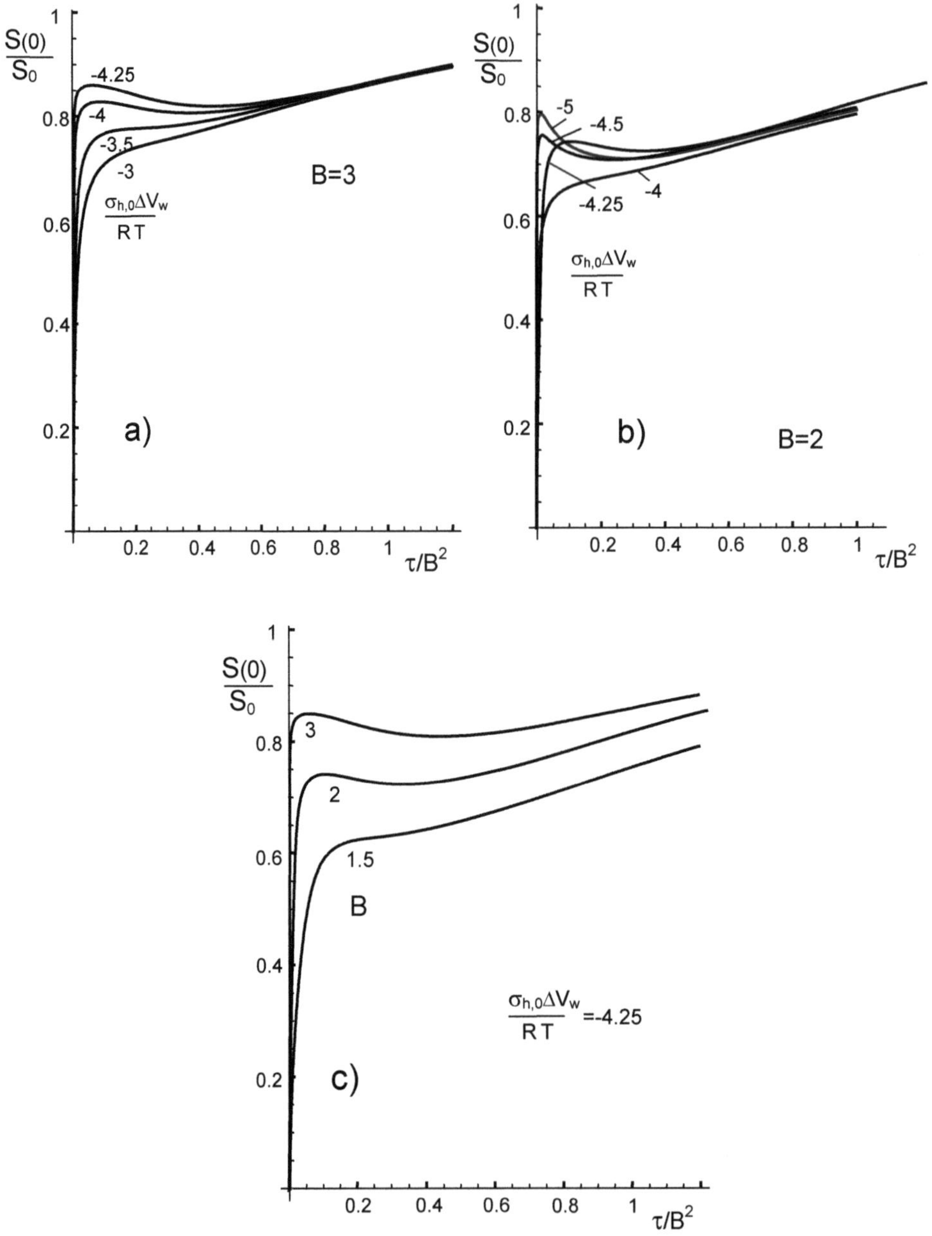

Fig. 1.15 Hydroxyl concentration at the surface, a) and b) effect of swelling stress, c) effect of the Biot number.

1.5 Water uptake

The uptake of molecular water, m_C, is obtained from eqs.(1.4.1a) with $\overline{C}(t)$ from (1.4.6)

$$m_C = \overline{C}(t) \times 2W = m_{C,0}\left(1 - \sum_{n=1}^{\infty} \frac{2B^2}{\alpha_n^2[B(B+1)+\alpha_n^2]} \exp\left[-\alpha_n^2 \frac{\tau}{B^2}\right]\right) \tag{1.5.1}$$

where the asymptotically reached value of molecular water uptake is denoted as

$$m_{C,0} = 2WC_0 \tag{1.5.2}$$

Equation (1.5.1) holds in the absence of stresses. In this case the hydroxyl content is

$$m_S(t) = k_{1,0} m_C(t) \tag{1.5.3}$$

In *presence of swelling stresses* the equilibrium constant is

$$k_1 = k_{1,0} \exp\left[\frac{\sigma_h \Delta\overline{V}}{RT}\right] \tag{1.5.4}$$

with the hydrostatic stress in *first order*

$$\sigma_h = -\frac{2\kappa E}{9(1-\nu)} k_{1,0}\, C \tag{1.5.5}$$

An approximate analytical solution can be obtained following the general procedure usual in perturbation theory. If we consider the equilibrium constant k_1 as the disturbance parameter, perturbation theory suggests to solve the problem in the absence of stresses, i.e. for k_1=$k_{1,0}$ =const. and to insert the stress-dependent disturbance parameter into this solution. The analytical solution of the undisturbed problem in terms of molecular water is $C(z,t)$ given by eq.(1.4.1-1.4.4).

The *first-order* solution for the hydroxyl water uptake under swelling stresses then reads

$$m_S(t) = k_{1,0} \int_0^{2W} C(z,t) \exp\left[\lambda \frac{C(z,t) - \overline{C}}{C_0}\right] dz \tag{1.5.6}$$

with the abbreviation λ defined by eq.(3.1.14).

The effect of the disappearing stresses due to the exponential term $C(z,t)-\overline{C}_0 \rightarrow 0$ for $t \rightarrow \infty$ results in a reduction of the hydroxyl uptake as is shown in [1.24].

References

1.1 S. M. Wiederhorn, F. Yi, D. LaVan, T. Fett, M.J. Hoffmann, Volume Expansion caused by Water Penetration into Silica Glass, J. Am. Ceram. Soc. **98** (2015), 78-87.

1.2 Doremus, R.H., Diffusion of water in silica glass, J. Mater. Res., **10** (1995), 2379-2389.

1.3 A. Zouine, O. Dersch, G. Walter and F. Rauch, "Diffusivity and solubility of water in silica glass in the temperature range 23-200°C," *Phys. Chem. Glass: Eur. J. Glass Sci and Tech. Pt. B*, **48** [2] (2007), 85-91.

1.4 Davis, K.M., Tomozawa, M., Water diffusion into silica glass: structural changes in silica glass and their effect on water solubility and diffusivity, J. Non-Cryst. Sol. **185** (1995), 203-220.

1.5 M. Helmich and F. Rauch, "On the mechanism of diffusion of water in silica glass," *Glastech. Ber.* **66** [8] (1993), 195-200.

1.6 Oehler, A., Tomozawa, M., Water diffusion into silica glass at a low temperature under high water vapor pressure, J. Non-Cryst. Sol. **347** (2004) 211-219.

1.7 R.H. Doremus, Diffusion of Reactive Molecules in Solids and Melts, Wiley, 2002, NewYork.

1.8 Carslaw, H.S., Jaeger, J.C. Conduction of heat in solids, 2nd ed. 1959, Oxford Press, London.

1.9 Wakabayashi, H., Tomozawa, M., Diffusion of water into silica glass at low temperature, J. Am. Ceram. Soc. **72** (1989), 1850-55.

1.10 H. Le Chatelier, *C.R. Acad. Sci. Paris* **99**(1884), 786

1.11 S.M. Wiederhorn, G. Rizzi, S. Wagner, M.J. Hoffmann, T. Fett, Diffusion of water in silica glass in the absence of stresses, J. Am. Ceram. Soc. **100** (2017), 3895–3902.

1.12 *Mathematica*, Wolfram Research, Champaign, USA.

1.13 G. Schell, T. Fett, Mass transfer coefficients for water at silica surfaces - Additional results, Scientific Working Paper SWP **73**, 2017, ISSN: 2194-1629, Karlsruhe, KIT.

1.14 H. Li and M. Tomozawa, Mechanical strength increase of abraded silica glass by high pressure water vapor treatment, J. Non-Cryst. Sol, **168** 287-291 (1994).

1.15 G. Perera, R.H. Doremus, J. Am. Ceram. Soc. **74** (1991), 1269-74.

1.16 P.M. Dove, N. Han, A.F. Wallace, J.J. De Yoreo, Kinetics of amorphous silica dissolution and the paradox of the silica polymorphs, PNAS, 105, [29] (2008), 9903-9908.

1.17 T.S. Mahadevan, S.H. Garofalini, Dissociative chemisorption of water onto silica surfaces and formation of hydronium ions, J. Phys. Chem. C **112**(2008), 1507-1515.

1.18 Feuston, B.P., Garofalini, S.H., Water induced relaxation of the vitreous silica surface, J. Appl. Phys. **68**(1990), 4830-4836.

1.19 Leed, E.A., Pantano, C.G., Computer modelling of water adsorption on silica and silicate glass fracture surfaces, J. Non-Cryst. Sol. **325**(2003), 48-60.

1.20 Ma, Y., Foster, A.S., Nieminen, R.M., Reactions and clustering of water with silica surface, J. Chem. Phys. **122** 144709 (2005).

1.21 S. M. Wiederhorn, G. Rizzi, S. Wagner, M. J. Hoffmann, and T. Fett, Stress-Enhanced Swelling of Silica: Effect on Strength, J. Am. Ceram. Soc. **99** [9] 2956–2963 (2016)

1.22 P.J. Lezzi, Q.R. Xiao, M. Tomozawa, T.A. Blanchet, C.R. Kurkjian, Strength increase of silica glass fibers by surface stress relaxation: A new mechanical strengthening method, J. of Non-Crystalline Solids **379** (2013) 95–106.

1.23. N.P. Bansal and R.H. Doremus, *Handbook of Glass Properties*, Academic Press, Inc. Orlando, FL (1986), pp. 25.

1.24 T. Fett, G. Schell, Water diffusion in thin silica sheets, Effect of swelling stress on diffusivity, Scientific Working Papers **72**, 2017, ISSN: 2194-1629, Karlsruhe, KIT.

2 Swelling in the absence of external stresses

2.1 Swelling and swelling stresses

2.1.1 Qualitative indications in literature

Clear evidence has been reported in the literature for volume swelling, ε_v, measured directly via dimension changes and indirectly via stress generation within the silica glass. The swelling behavior has been determined by length [2.1] and curvature measurements on glass bars [2.2] and by X-ray diffraction strain measurements [2.3] on silica surfaces. Gorbacheva and Zaoints [2.1] soaked fused silica prisms (25×5×5 mm) in water at 80 °C for 20 months. They found that the measured length had increased by 0.17% during this treatment. Since only a thin surface layer of water penetrated glass, ≈ 5 µm, could have been generated during the water storage, the suppressed linear strain in the surface must be clearly larger than 0.17%.

2.1.2 Swelling strains from density measurements

In the past, the authors showed in a couple of papers the principle effects of volume swelling by hydroxyl generation in silica on mechanics and fracture mechanics properties [2.4, 2.5, 2.6].

A swelling effect in water-containing silica at high temperatures was early reported by Brückner [2.7, 2.8], Shackelford [2.9] and Shelby [2.10]. These authors showed that the density decreased by reaction (1.1.1) for a fictive temperature of T_f=1100°C according to

$$\frac{\Delta\rho}{\rho_0} = -\chi C_w\,,\quad C_w = \frac{m_w}{m_{glass}+m_w} \cong \frac{m_w}{m_{glass}} \tag{2.1.1}$$

where C_w is the weight fraction of water. The literature data for a fictive temperature of T_f=1100°C are shown in Fig. 2.1 by the open symbols. The individual data for Brückner and Shelby were taken from the compilation by Shelby [2.10]. The data of Shackelford for fictive temperatures T_f=1000°C, T_f=1100°C and T_f=1200°C were directly taken from the original paper [2.9] transforming the "water" content (wt% OH) in water content (wt% H_2O) using the factor (18/17)/2 where 18/17 reflects the different molar weights of OH and H_2O.

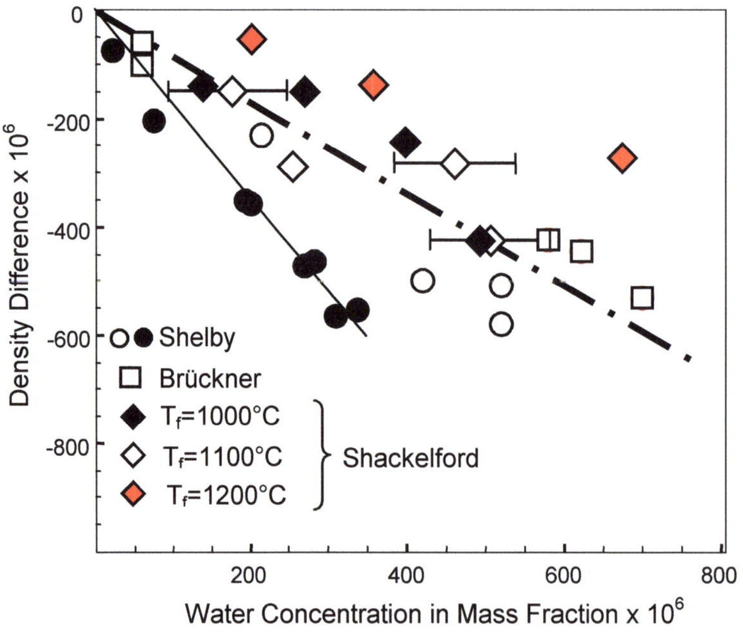

Fig. 2.1 Effect of water concentration on density of vitreous silica, results by Shelby [2.10], Brückner [2.7, 2.8], and Shackelford [2.9]. The dash-dotted line represents the value given in (2.1.2). The open symbols are for a fictive temperature of T_f=1100°C, the black symbols for T_f=1000°C and the red ones for T_f=1200°C. The circles represent results by Shelby [2.10] with the water produced in silica saturated with hydrogen and then gamma-irradiated.

By a least-squares fit of the data points for T_f=1100°C and 1000° exclusively, we obtain

$$\chi = 0.836\,[0.739, 0.933] \tag{2.1.2}$$

with the 95%-Confidence Interval in brackets. The dependency given by this value is indicated in Fig. 2.1 as the dash-dotted line. In our former papers [2.11, 2.4-2.6] we used the value of χ=0.84 as was obtained simply as the average of the 3 values given in [2.10].

From the definition of the density as the quotient of mass $m=m_{glass}+m_w$ and volume V, $\rho=m/V$, it follows for the volume swelling strain ε_v

$$V = \frac{m}{\rho} \Rightarrow \varepsilon_v = \frac{\Delta V}{V} = \frac{\Delta m}{m} - \frac{\Delta \rho}{\rho} \tag{2.1.3}$$

Since only the water content m_w can change during water soaking, the mass change is

$$\Delta m = m_w = C_w m \tag{2.1.4}$$

for $m_w << m_{glass}$. Consequently, the volume swelling strain ε_v simply results from eqs.(2.1.1-2.1.4) as

$$\varepsilon_v = (1+\chi) C_w \tag{2.1.5a}$$

In the following, let us consider the case of swelling behavior for both the ≡SiOH and the molecular water that is present at low temperatures <500°C. The total swelling strain ε_v is then composed of the individual contributions of the hydroxyl, ε_S, and the molecular water, ε_C, according to a molar "rule of mixture":

$$\varepsilon_v = \frac{S/2}{C+S/2}\varepsilon_S + \frac{C}{C+S/2}\varepsilon_C \tag{2.1.5b}$$

So far, no reliable information on volume change is available for the strain ε_C by the molecular water species. We believe that the molecular water in the glass is located in holes of molecular size in the silica glass network [2.12, 2.11], where the water may not contribute to the volume change, i.e. we *assume* in the following considerations that ε_C=0. On the other hand, it is well known that molecular water at high temperatures of 1100°C is nearly absent. Under these circumstances, the volume swelling in [2.7-2.10] is due to hydroxyl generation by the reaction (1.1.1), i.e. $\varepsilon_v=\varepsilon_S$. This may be confirmed and if necessary improved by future experiments of Molecular Dynamic (MD) computations. For this topic see also the Outlook-Section.

2.1.3 Transformation into hydroxyl concentrations and swelling strains

In general C_w has to be converted to mole fraction of water, x_{H2O}, in the glass via

$$x_{H2O} = \frac{M_{glass}}{m_w} C_w = (60/18) C_w \tag{2.1.6}$$

Here we used the molar mass for water m_w =18g/mol and as an approximation M_{glass}= 60g/mol for the molecular weight of the glass. For every water molecule that reacts with the glass, two SiOH are formed, therefore:

$$x_{SiOH} = 2 x_{H2O} \tag{2.1.7}$$

Combining (2.1.6) and (2.1.7) yields

$$x_{SiOH} = 2 \frac{60}{18} C_w \tag{2.1.8}$$

By assumption, the volume expansion is just due to the ≡SiOH formation. The hydroxyl *mass concentration* S is

$$S = x_{SiOH} \frac{m_{SiOH}}{M_{glass}} = x_{SiOH} \frac{17}{60} \tag{2.1.9}$$

Consequently, the volume swelling strain in terms of the water concentration C_w is

$$\varepsilon_v = (1+\chi) C_w = 1.836 \times C_w \ \ [1.739 \times C_w, 1.933 \times C_w] \tag{2.1.10}$$

Here it should be emphasized once more that this dependency holds for *small* water concentrations. This relation reads in terms of hydroxyl mass concentration S [2.13]:

$$\varepsilon_v = \tfrac{18}{17} (1+\chi) \frac{S}{2} = \kappa \times S \tag{2.1.11}$$

Then

$$\kappa = 0.97 \ \ [0.92, 1.02] \tag{2.1.12}$$

In (2.1.12) the numbers in brackets represent the 95% confidence interval. The value of κ=0.97 is recommended to be used for “best” prediction, the lower boundary value of the CI, κ=0.92, for “conservative” predictions. Our coefficient used so far [2.11], $\kappa=0.92$, is located at the lower boundary of the 95%-CI.

The data by Shackelford for T_f=1000°C represented by the black solid symbols fit very well to the data for T_f=1100°C. Therefore, we included these data in the analysis with the result of

$$\kappa = 0.965 \;\; [0.916, 1.014] \tag{2.1.13}$$

that is practically identical with the value of (2.1.12). Including all data for the three fictive temperatures gives

$$\kappa = 0.92 \;\; [0.86, 0.984] \tag{2.1.13a}$$

To our actual knowledge we suggest use of κ=0.97. Having in mind the limited accuracy of water concentrations (in the Shackelford analysis ±75 ppm H_2O) the equation for water concentrations in terms of mass units seems to be sufficiently accurate also for molar units because the mole masses deviate not strongly, i.e.

$$\frac{m_w}{m_{SiOH}} = \frac{18 \text{ g/mol}}{17 \text{ g/mol}} = 1.058 \approx 1 \tag{2.1.14}$$

2.1.4 Results by Shelby on hydrogen infiltrated and gamma-irradiated silica

Shelby [2.10] generated water in the glass on a further way. He saturated the glass with hydrogen H_2 under gas pressures up to 70 kPa. The specimens were then gamma-irradiated to doses of 1 to 15 Grads. In this way, SiOH was directly produced in the bulk material. The resulting density reduction by this procedure is shown in Fig. 2.1 by the solid circles. These results do strongly deviate by roughly a factor of two from all the other data. The regression analysis according to [2.14] for these data gives

$$\frac{\Delta\rho}{\rho} = -\chi\, C_w\,, \quad \chi = 1.735\,[1.60, 1.87] \tag{2.1.15}$$

For the computation of the volume strain it has to be taken into account that the resulting water has not been introduced from the environment, but was generated in situ with the oxygens taken from the SiO_2 network.

Unfortunately it is not clearly visible in [2.10] whether the initial density was measured before or after saturation by hydrogen. Therefore, we will perform two different evaluations as performed in [2.15]:

Assumption 1: The initial density was measured *before* hydrogen infiltration. Consequently, the mass change is 2/18 C_w since 2 hydrogen atoms are needed for each water molecule.

$$\Delta m = \tfrac{1}{9} C_w m \tag{2.1.16}$$

In this case, the volume strain results as

$$\varepsilon_v = (\tfrac{1}{9} + \chi) C_w \tag{2.2.17}$$

and according to (2.1.11) we obtain

$$\varepsilon_v = \tfrac{18}{17}(\tfrac{1}{9} + \chi)\frac{S}{2} = \kappa \times S \tag{2.1.18}$$

Introducing the result of eq.(2.1.15) yields

$$\kappa = 0.977 \ [0.906, 1.049] \tag{2.1.19}$$

Assumption 2: The initial density was measured *after* the hydrogen infiltration. In this case it is simply Δm=0

$$\varepsilon_v = \tfrac{18}{17}\chi\frac{S}{2} = \kappa \times S \tag{2.1.20}$$

From this it yields

$$\kappa = 0.92 \ [0.847, 0.99] \tag{2.1.21}$$

Within the 95%-confidence intervals, also the results of (2.1.19) and (2.1.21) are in reasonable agreement with the results from heat-treated specimens.

Finally, the volume increase ε_v is plotted in Fig. 2.2a as a function of the water concentration C_w and in Fig. 2.2b versus the hydroxyl concentration S for all measurements in Fig. 2.1. The symbols in Fig. 2.2 are the same as given in Fig. 2.1. The straight line, introduced in Fig. 2.2b, represents the coefficient κ=0.97 according to (2.1.13). The good agreement with all the data is evident.

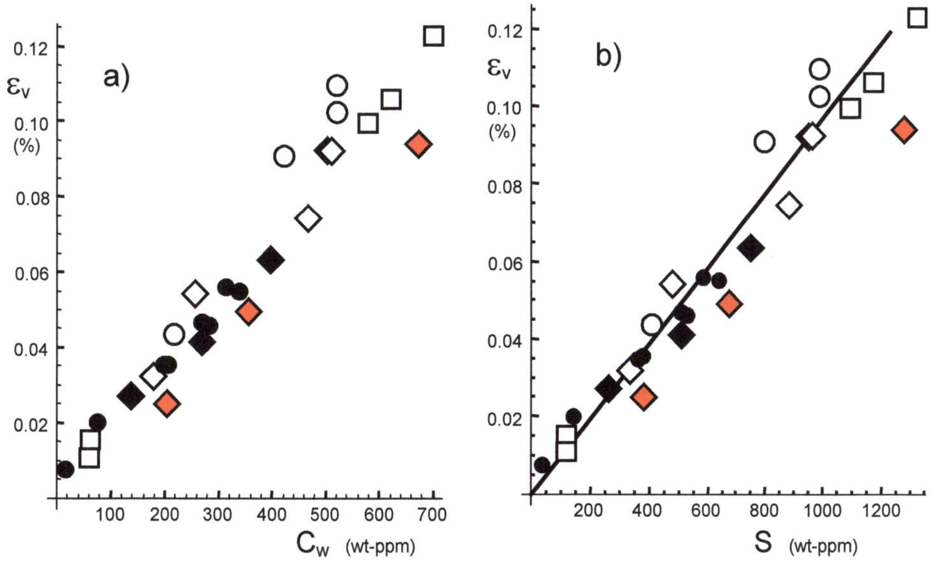

Fig. 2.2 a) Volume expansion strain vs. water concentration, b) expansion vs. hydroxyl concentration S, (symbols as in Fig. 2.1), line: eqs.(2.1.20) and (2.1.21), from [2.14, 2.15].

2.1.5 Swelling stresses under isotropic swelling

The effect of the water-silica reaction is the generation of *swelling strains. Swelling stresses* are a consequence of the mechanical boundary conditions. A volume element in a thick plate that undergoes swelling cannot freely expand. If the diffusion zone is small compared to the component dimensions, expansion is completely prevented in the plane of the surface and can only take place normal to the surface plane.

At a free surface, the stress state is *plane stress* and, consequently, also stresses caused by swelling are equi-biaxial (σ_x=0)

$$\sigma_{sw,y} = \sigma_{sw,z} = -\frac{\varepsilon_v E}{3(1-\nu)} \tag{2.1.22}$$

In (2.1.22) E is Young's modulus and ν is Poisson's ratio. Consequently, the hydrostatic stress term reads

$$\sigma_{sw,h} = \tfrac{1}{3}(\sigma_{sw,y} + \sigma_{sw,z}) = -\frac{2\,\varepsilon_v\, E}{9(1-\nu)} \tag{2.1.23}$$

Introducing the proportionality (2.1.11) in (2.1.22) gives

$$\sigma_{sw,h} = -\frac{2\,\kappa E}{9(1-\nu)} S = -18.7\ \mathrm{GPa} \times S \tag{2.1.24}$$

2.2 Swelling from measurement of disk curvature

2.2.1 Evaluation of curvature measurements

Swelling stresses were derived in [2.11] by measuring the change of the curvature of silica disks after hot-water soaking. At the heart of this study is an experiment in which water is diffused into one side of thin vitreous silica disks. This technique has been used earlier to determine the volume change caused by the penetration of water into silica coatings on silicon wafers [2.2]. The consequent volume expansion at the glass surface results in a bending of the disks, which are no longer flat after exposure, but concave-shaped [2.16]. By measuring the radius of curvature at the surfaces of the curved disks and by understanding the mechanism of water migration, the surface stress can be calculated. The technique can also be used to calculate the diffusivity of water in silica glass.

Two different experimental procedures were applied in [2.11] for tests at 90°C and 200°C.

90°C: Silica disks of W=250µm thickness were coated on one side by e-beam evaporation with a 3 nm bond-coat of titanium, followed by 10 nm of gold. Then the initial curvature, R_0, was measured using the wafer-curvature, laser-scanning technique. Disks were exposed to water at 90 °C for periods ranging from two to six weeks. One specimen series was immersed in water, a second test series suspended over the water – the uncoated side facing the water, the coated side facing the relatively dry furnace air.

A plate is considered with all dimensions very much larger than the water diffusion zone. The distribution of the swelling stresses is given by

$$\sigma_{sw} = \sigma_0 \operatorname{erfc}\left(\frac{z}{2b}\right) \tag{2.2.1}$$

with

$$b = \sqrt{Dt} \tag{2.2.2}$$

The bending moment caused by a thin swelling layer is then given as

$$M_b = \frac{W}{2}\sigma_0 \underbrace{\int_0^\infty \operatorname{erfc}\left(\frac{z}{2b}\right) dz}_{\frac{2b}{\sqrt{\pi}}} = W\sigma_0 \frac{b}{\sqrt{\pi}} \tag{2.2.3}$$

The measurable change of the curvature by swelling is (see e.g. Timoshenko and Goodier, Theory of Elasticity, page 290)

$$\Delta\frac{1}{R} = \frac{12(1-\nu)}{EW^3} M_b \tag{2.2.4}$$

In [2.11] the diffusion parameter b was computed with the diffusivity given by Zouine et al. [2.17]. Using the surface water concentration reported by Zouine et al. [2.17] together with the S/C-ratio from IR-measurements in [2.11], the surface swelling stress σ_0 was obtained via

$$\sigma_0 = -\frac{\kappa E}{3(1-\nu)} S \tag{2.2.5}$$

The bending moments computed via eq.(2.2.3) are compared in Fig. 2.3 with the experimental results obtained from (2.2.4). Figure 2.3 shows the measured and computed moments plotted versus square root of time.

The prediction made with the surface concentration from [2.17] is entered as the triangle. In a second attempt the total water uptake m_{Cw} was used as measured by IR as well as the

S/*C* ratio also obtained from IR. As shown in [2.11] it holds for the generated bending moment expressed in terms of the hydroxyl concentration *S*

$$M_b = -\kappa \frac{W}{2} \frac{E}{3(1-\nu)} \cdot m_S \qquad (2.2.6)$$

where m_S is the uptake of the hydroxyl water from one side.

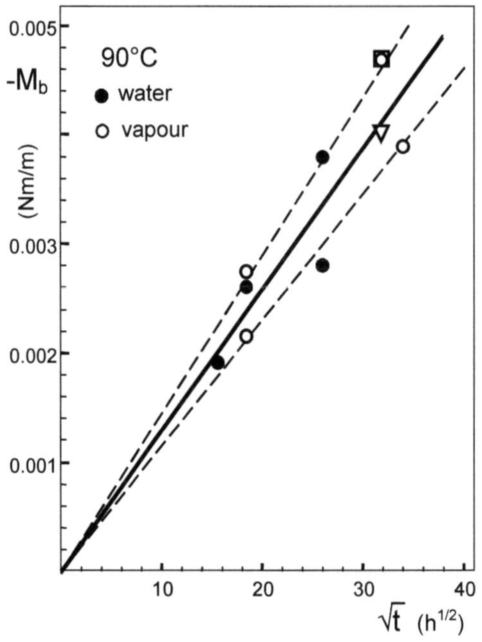

Fig. 2.3 Measured bending moments (via disk curvature) versus time (circles) for 90°C. The predicted moment obtained from *IR* data is given by the square. The triangle is determined via the water concentration data from Zouine *et al.* [2.17].

The result from this equation is plotted in Fig. 2.3 as the square. The prediction using the surface water concentration by Zouine et al. [2.17] (triangle) is in very good agreement with the curvature measurements. Even though the square lies just outside of the confidence limits, these limits are for the mean value of M_b and not the individual data points. A certain fraction of the data should lie outside of the confidence band. The two predictions indicate support for the assumption that only the ≡SiOH species contribute to the volume expansion during soaking.

200°C: At this temperature it was not possible to prevent water entrance sufficiently by a gold layer. Therefore the bare silica disk was heat-treated on both sides and then stepwise etched from one side in a buffered HF solution. After each etching step the change of curvature was determined resulting in the released bending moment versus total removed surface layer as plotted in Fig. 2.4a for 2h and 20h soaking times at 196°C.

In order to avoid the complicated equations in integral form used in [2.11], the procedure may be explained here alternatively in differential form.

The bending moment per unit length caused by a thin surface layer of stresses σ in a plate of thickness *W* is given by

$$M_b = \frac{W}{2}\int_0^{\infty} \sigma(z)\, dz \qquad (2.2.7)$$

If the surface is removed by an amount of *d*, the remaining moment after a total surface removal (single etching depths accumulated) is

$$M_b(d) = \frac{W}{2}\int_d^{\infty} \sigma(z)\, dz \qquad (2.2.8)$$

and the change of moment

$$\Delta M_b(d) = \frac{W}{2}\int_0^{d} \sigma(z)\, dz \qquad (2.2.9)$$

The left-hand side of (2.2.9) is known from measurements of curvature by using

$$\Delta M_b(d) = \frac{EW^3}{12(1-\nu)}\left(\frac{1}{R} - \frac{1}{R_0}\right) \qquad (2.2.10)$$

where R is the actual and R_0 the initial curvature.

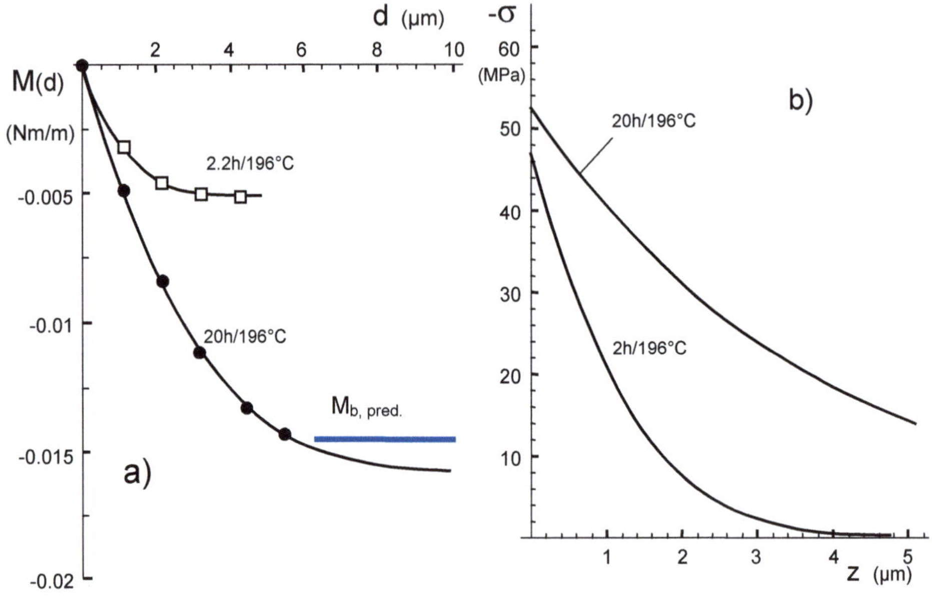

Fig. 2.4 a) Change of the bending moment during stepwise etching one surface, b) swelling stresses.

Equation (2.2.9) is an integral equation with respect to the unknown stress distribution $\sigma(z)$. Since the integrand does not depend on d explicitly, its solution can simply be obtained by taking the derivative with respect to the removed layer thickness d

$$\sigma\big|_{z=d} = \frac{2}{W}\frac{\partial \Delta M_b}{\partial d} \tag{2.2.11}$$

This equation holds for the case that the removed layer-thickness is small compared with the thickness of the disk, i.e. for $d \ll W$. The measured change of the bending moment is plotted in Fig. 2.4a. In the procedure described in [2.11] the swelling stresses were first represented by appropriate erfc-type functions with unknown parameters and then integrated according to eq.(2.2.7). The resulting bending moments were then fitted to the measured data resulting in the best set of parameters for the stress distribution $\sigma(z)$. In the procedure based on eq.(2.2.11) the measured data were transformed into stresses and these were then "smoothed" by fitting with erfc-type functions. For taking the derivative, the measured data were interpolated by cubic splines using the Mathematica subroutine *Interpolation* [2.18]. Then the splines were differentiated with the result shown in Fig. 2.4b.

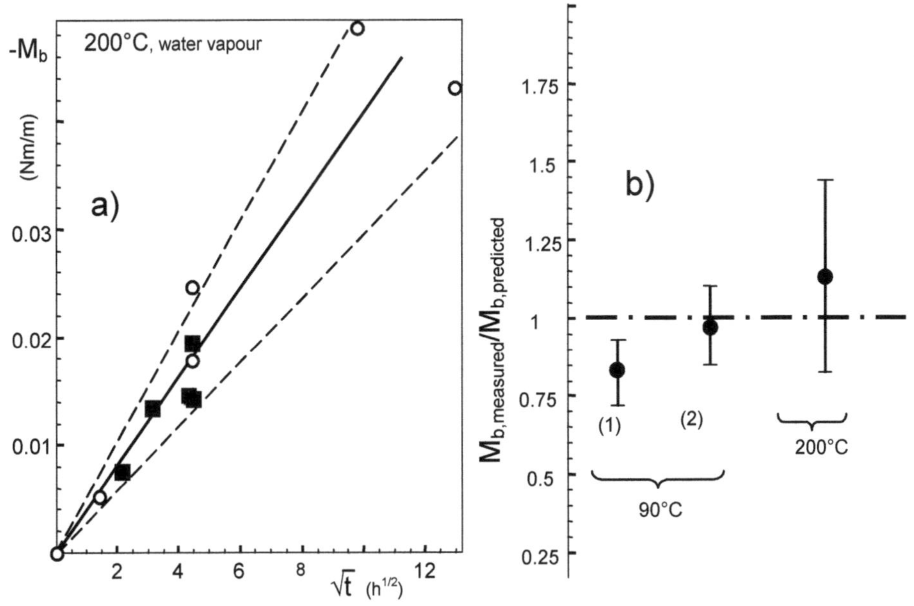

Fig. 2.5 a) Comparison of measurements (circles) and predictions (squares) based on measured IR-data and data reported by Zouine *et al.* [2.17], and Helmich and Rauch [2.19]. The dashed lines represent the 95% error limits of measurements; b) Ratio of measured and predicted bending moments, bars represent ±1 SD, circles: predicted with κ=0.97, eq.(2.1.12), (1) based on IR-measurements by Wiederhorn et al. [2.11], (2) based on data by Zouine et al. [2.17].

Fig. 2.4b gives the stress distribution obtained from the curves of Fig. 2.4a. Finally, Fig. 2.5a represents the total bending moment released by removing the swelling layer completely. This value is directly visible for the case of 2h soaking time in Fig. 2.4a. In cases where a plateau value is not yet visible in Fig. 2.4a, the value can be computed by eq.(2.2.7) using the erfc-type functions from the fitting procedure as the integrand.

All predicted data in Figs. 2.3 and 2.5a (squares and triangle) are within the 95% error limits supporting again the assumption that only the ≡SiOH contributes to the volume expansion. This can be seen from Fig. 2.5b. In this figure, the ratio of the measured to the predicted bending moments is plotted. The symbols represent the predictions made in [2.11], the bars represent ±1 Standard Deviation. The results of Fig. 2.5b are compiled also in Table 2.1. (1) gives the bending moment for water-soaked silica disks at 90°C by Wiederhorn et al. [2.11] together with the prediction based on the water content obtained via IR-measurement. The numbers in brackets represent ±1 standard deviation (S.D.).

In (2) the same experimental data are compared with the predictions using water concentrations at the surface as were reported by Zouine et al. [2.17].

	Moment (Nm/m)	predicted	measured	ratio
(1)	$-M_b$(90°C)	0.0047	0.00413 [0.00363, 0.00463]	0.878 [0.772, 0.985]
(2)	$-M_b$ (90°C) [2.11]	0.0040		1.032 [0.908, 1.16]
(3)	$-M_b$ (200°C) [2.11]	0.0145	0.0173 [0.0128, 0.0217]	1.19 [0.882, 1.496]

Table 2.1 Comparison of predicted and measured bending moments from [2.11].

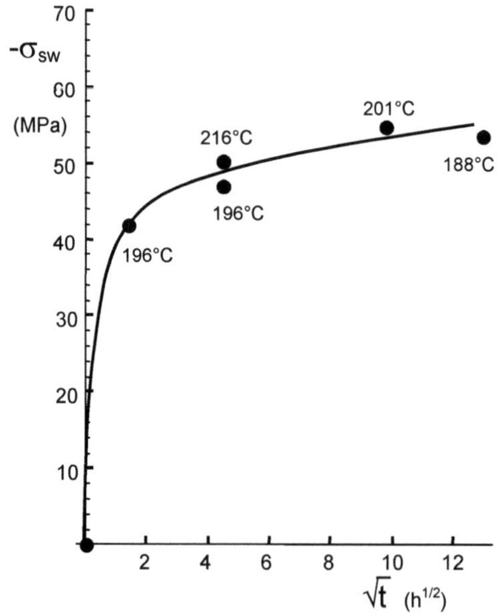

Fig. 2.6 Maximum swelling stresses, results from [2.11] for temperatures in the range of 188°C≤θ≤216°C.

Finally, Fig. 2.6 summarizes the surface swelling stresses from [2.11] plotted as a function of heat-treatment time. In context with the data scatter, it should be mentioned that the individual data points represent slightly different soaking temperatures as indicated at the individual data points.

2.2.2 Conclusions from curvature measurements

In literature a different interpretation of the mechanical consequences of the silica-water reaction (1.1.1) is actually discussed by Tomozawa and co-workers [2.20], [2.21]. In these papers it is suggested that the mechanical response on water entrance would be an instantaneous and complete *relaxation* of stresses pre-existing at a silica surface. This hypothetical mechanism is claimed to occur even at 200°C. This makes it possible to have a look on such a mechanism with regard to the bending tests at this temperature [2.11].

Our ***conclusions*** from the point of the results on disks are:

- Before the tests, the silica disks were etched in 5% volume fraction HF in order to remove 5-10µm of glass, which might have contained a diffusion layer of water at the surface or might have residual stresses due to polishing. Consequently, stress relaxation could not be the reason for curvature since no stresses to relax were present.
- We have to conclude that the mechanical effect of the reaction (1.1.1) is an *active process* that generates strains and depending on the mechanical boundary conditions stresses. It is not a *passive* effect like stress relaxation that needs pre-existing stresses.
- On the other hand, Fig. 2.5b clearly shows that the predictions and measurements of curvature (or bending moments) are within 0.85-1.20 of the ideal agreement (dash-dotted line =1). We have to state that there is no free space for any further stress-generating effect. This confirms the assumption that the water species causing swelling strains is exclusively the hydroxyl S even at rather low temperatures.

In contrast to a suggested relaxation effect, it has furthermore to be emphasized that the swelling mechanism is already proved by the density reduction as the consequence of an increase of the water content [2.8-2.10].

Moreover that relaxation is not present at 200°C or at least negligible. If relaxation would be present at 200°C, the generated stresses in the disks causing the observed curvature must have relaxed instantaneously and of course not any curvature could result as observed.

2.2.3 Equilibrium constant

The equilibrium constant k_1 according to eq.(1.1.2) is plotted in Fig. 2.7a. The solid circles represent the data obtained in [2.11] and the open ones are results from literature published by Zouine [2.17], Davis and Tomozawa [2.22, 2.23], and Oehler and Tomozawa [2.24]. The data follow an Arrhenius curve, *i.e.*, a straight line in Fig. 2.7a that is expressed by eq.(1.1.3). The activation energy of the least squares fit line is Q=10.8±1.1 kJ/mol. The parabolic curves on each side of the straight line are the 95% confidence limits of the fit.

2.3 Observation on etching rates

In the etching procedure in hydrofluoric acid/sulfuric acid solutions at room temperature; 2 N NaOH at 80 °C, we found as an interesting detail that the etching rate decreased with increasing swelling stresses. Results are shown in Fig. 2.7b plotted versus the swelling stress for disks that were soaked for 2 h at about 200°C under saturation water vapour pressure of 1550 kPa. The open circles show results at 216°C, the solid circles stand for a test at 196°C.

From Fig. 2.7b we found that the etching rate decreased with increasing compressive swelling stresses. A fit over all data according to

$$\frac{d\delta}{dt} = a_0 \exp[a_1 \sigma_{sw}] \tag{2.3.1}$$

($d\delta$ is the removed layer thickness during the time interval dt) resulted in the parameters

a_0=0.1179 µm/min [0.1154, 0.1203], a_1=0.0032 MPa^{-1} [0.0024, 0.0041]

with the 90% Confidence Intervals in brackets. Since the hydrostatic swelling stress at the surface is 2/3rd of the equi-biaxial stress, a representation via an Arrhenius relation gives the stress dependence of the etching rate by

$$\frac{d\delta}{dt} = a_0 \exp\left(\frac{\sigma_{sw,h} V}{RT}\right) \tag{2.3.2}$$

with an activation volume V,

$$V = 11.7\ \text{cm}^3/\text{mol}\ \ [8.8,\ 15]$$

again with 90% Confidence Interval.

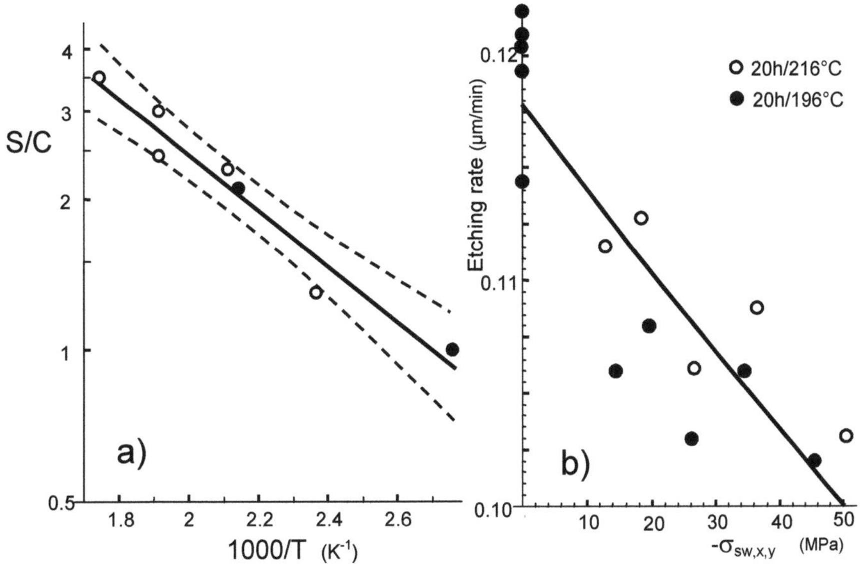

Fig. 2.7 a) Plot of S/C as a function of $1/T$. The solid circles are from [2.11] via IR measurements on disks exposed to saturated steam at 200 °C and 90 °C. The open circles are IR measurements made by other authors, summarized in Zouine *et al.*[2.17], b) etching rate as a function of the average stress in the etching intervals for silica disks hot water soaked at about 200°C for 20 h.

In contrast to the Charles/Hillig theory Tomozawa and co-workers [2.25, 2.26] report negative activation volumes. This means that under compressive stresses the corrosion rates would be larger than under tension. This is not necessarily in contradiction with the results in Fig. 2.7b. It may be possible that the etching rate in completely stress-free specimens containing high hydroxyl concentrations would strongly decrease and overcompensate a possible inverse stress effect. Soaking tests after very long times would result in a homogeneously distributed S concentration over the whole specimen with disappearing swelling stresses. The only conclusion from the etching rate measurements is so far that the higher the S-concentration is, the smaller are the etching rates.

2.4 Volume swelling and shrinking in other glasses

In the previous Sections, only silica was considered. In this Section, we will look on other glasses, too. Density measurements on several glass compositions were reported by Scholze [2.27]. The materials were: (B) B_2O_3-glass, (K) K_2O-SiO_2-glass (20-80 mol-%), (Ca) Na_2O-CaO-SiO_2-glass (16-10-74 mol-%), (Al) Li_2O-Al_2O_3-SiO_2-glass (20-5-75 mol-%), and (Si) silica.

The relative density changes due to water content C_w are given in Fig. 2.8a. Whereas for silica a non-linear behaviour $\Delta\rho(C_w)$ is reported in [2.27], for all other glasses straight-line behaviour was found. It has to be taken into account that the data scatter for silica is rather large as the data reported by Shelby [2.10] show which are introduced in red. The straight line represents the average of data for *as-received* water in silica measured by several authors and quoted by Shelby [2.10]. The circles are results by Brückner [2.8] measured by *water removal* (also quoted in [2.10]).

Since the studies resulting in the red line ([2.7-2.10]) showed reasonable agreement, we have used this curve for our computations. For a discussion of the trustworthiness of the different results for silica, see Shelby [2.10].

Glass	Composition (mol-%)	(wt-%)	ρ_0 (g/cm^3)	$\frac{\Delta\rho}{\rho_0 C_w}$	ε_v/C_w, eq.(2.1.5a)
B	B_2O_3-glass	-	1.83	6.3	−5.3
K	K_2O-SiO_2-glass (20-80)	28.2-71.8	2.38	1.41	−0.41
Ca	Na_2O-CaO-SiO_2-glass (16-10-74)	16.6-9.4-74.1	2.49	0.46	0.54
Al	Li_2O-Al_2O_3-SiO_2-glass (20-5-75)	10.7-9.1-80.2	2.33	0.37	0.63
Si	SiO_2	-	2.2	−0.84	1.84

Table 2.2 Swelling and shrinking strains derived from density curves given by Scholze [2.27] using eq.(2.1.5a).

Whereas silica shows a density decrease with increasing water content, the densities of the other glasses increased with increasing C_w. The relative density changes $\Delta\rho/\rho_0$ were transformed into volume strains ε_v according to eq.(2.1.3). The results are plotted in Fig. 2.8b and compiled in Table 2.2. While the B_2O_3-glass shows strong volume shrinking due to water uptake, silica shows clear volume expansion. The effects on the other glasses are rather small. Materials of Table 2.2 containing Ca and Al also swell. *K* shrinks slightly.

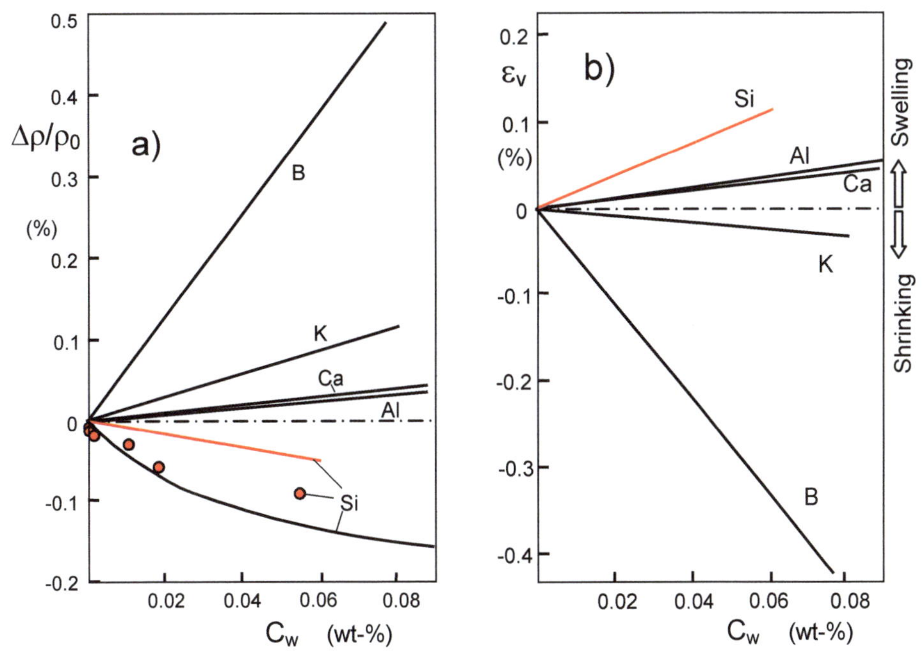

Fig. 2.8 a) Density change of glasses as a function of water content. Black curves given by Scholze [2.27], red line: average of data for as-received water in silica quoted by Shelby [2.10], circles: data from Brückner for water removal, also quoted in [2.10]. B: B_2O_3-glass, K: K_2O-SiO_2-glass (20-80 mol-%), Ca: Na_2O-CaO-SiO_2-glass (16-10-74 mol-%), Al: Li_2O-Al_2O_3-SiO_2-glass (20-5-75 mol-%); b) volume strains computed from Fig. 2.8a via eq.(2.1.3).

The swelling effect is evident also from quantum mechanical computations by Zhu et al. [2.28]. These authors made quantum mechanics computations on a SiO_2-nanorod of 36 SiO_2 atoms to which they added a single water molecule and determined the effect of the hydroxyl generation on stress-strain behaviour.

From their computations, a linear swelling stress of $\sigma_{sw,z} = -1.44$ GPa was found by the hydroxyl generation. From the axial modulus of 189 GPa it follows a *linear* swelling strain of $\varepsilon_{sw,z} \approx 0.8\%$. The numerical results must of course depend on the special SiO_2-arrangement (axially orientated rod) and the mechanical boundary conditions of the rod, nevertheless two points become evident:

a) The material must swell by the silica-water reaction.

b) The swelling stresses can become very large (in this special case −1.44 GPa).

References

2.1. M.I. Gorbacheva and R.M. Zaoints, "Moisture expansion of various crystalline and amorphous phases," *Steklo I Keramika*, **11** (1974), 18-19.

2.2. J. Thurn, "Water diffusion coefficient measurements in deposited silica coatings by the substrate curvature method," *J. Non-Cryst Solids*, **354** (2008), 5459-65.

2.3. W.M. Kuschke, H.D. Carstanjen, N. Pazarkas, D.W. Plachke, and E. Arzt, "Influence of water absorption by silicate glass on the strains in passivated Al conductor lines," *J. Electronic Mater.* **27** (1998), 853-857.

2.4 S.M. Wiederhorn, T. Fett, G. Rizzi, M. Hoffmann, J.-P. Guin, "Water Penetration – its Effect on the Strength and Toughness of Silica Glass," *Met. Mater. Trans. A*, **44** (2013) [3], 1164 -1174.

2.5 S.M. Wiederhorn, T. Fett, G. Rizzi, S. Fünfschilling, M.J. Hoffmann and J.-P. Guin, "Effect of Water Penetration on the Strength and Toughness of Silica Glass," *J. Am. Ceram. Soc.* **94** (2011) [S1] S196-S203.

2.6 T. Fett, G. Rizzi, M. Hoffmann, S. Wagner, and S.M. Wiederhorn, "Effect of Water on the inert Strength of Silica Glass: Role of Water Penetration," *J. Am. Ceram. Soc.* **95** (2012) [12], 3847-3853.

2.7 Brückner, R., "The structure-modifying influence of the hydroxyl content of vitreous silicas," Glastech. Ber. **43** (1970), 8-12.

2.8 Brückner, R., "Metastable equilibrium density of hydroxyl-free synthetic vitreous silica," J. Non-Cryst. Solids, **5** (1971), 281-5.

2.9 Shackelford, J.F., Masaryk, J.S., Fulrath, R.M., "Water Content, Fictive Temperature and Density Relations for Fused Silica," J. Am. Ceram. Soc. **53** (1970), 417.

2.10 Shelby, J.E., "Density of vitreous silica," J. Non-Cryst. **349** (2004), 331-336.

2.11 S. M. Wiederhorn, F. Yi, D. LaVan, T. Fett, M.J. Hoffmann, Volume Expansion caused by Water Penetration into Silica Glass, J. Am. Ceram. Soc. **98** (2015), 78-87.

2.12 J. F. Shackelford, "Gas Solubility in Glasses: Principles and Applications," Int. J. Appl. Glass Sci, **2** [2] (2011) 85–95.

2.13 S. M. Wiederhorn, M. J. Hoffmann, T. Fett, Swelling strains from density measurements, Scientific Working Papers **38**, 2015, KIT Scientific Publishing, Karlsruhe.

2.14 T. Fett, K.G. Schell, S.M. Wiederhorn, Swelling strains from gamma-irradiated silica - Evaluation of results by Shelby, Scientific Working Papers **103**, 2015, ISSN: 2194-1629, Karlsruhe, KIT.

2.15 T. Fett, K.G. Schell, S.M. Wiederhorn, Molar volume of SiOH estimated from swelling strains, International Journal of Applied Glass Science, **11** (2020), 608-611.

2.16. P.A. Flinn, "Principles and Applications of Wafer Curvature Techniques for Stress Measurements in Thin Films," *Mat. Res. Soc. Symp. Proc.* **130** (1989), 41-51.

2.17 A. Zouine, O. Dersch, G. Walter and F. Rauch, "Diffusivity and solubility of water in silica glass in the temperature range 23-200°C," *Phys. Chem. Glass: Eur. J. Glass Sci and Tech. Pt. B*, **48** [2] (2007), 85-91.

2.18 *Mathematica*, Wolfram Research, Champaign, IL, USA.

2.19 M. Helmich and F. Rauch, "On the mechanism of diffusion of water in silica glass," *Glastech. Ber.* **66** [8] (1993), 195-200.

2.20 M. Tomozawa, P.J. Lezzi, R.W. Hepburn, T. A. Blanchet, D. Cherniak, Surface Stress Relaxation and Resulting Residual Stress in Glass Fibers: A New Mechanical Strengthening Mechanism of Glasses, J. Non-Cryst. Solids **358**(2012), 2650.

2.21 P.J. Lezzi, Q.R. Xiao, M. Tomozawa, T.A. Blanchet, C.R. Kurkjian, Strength increase of silica glass fibers by surface stress relaxation: A new mechanical strengthening method, J. of Non-Crystalline Solids **379** (2013) 95–106.

2.22 Davis, K.M., Tomozawa, M., Water diffusion into silica glass: structural changes in silica glass and their effect on water solubility and diffusivity, J. Non-Cryst. Sol. **185** (1995), 203-220.

2.23 K.M. Davis, and M. Tomozawa, "An infrared spectroscopic study of water-related species in silica glasses," "J. Non-Cryst. Solids, **201** (1996), 177-198.

2.24 Oehler, A., Tomozawa, M., Water diffusion into silica glass at a low temperature under high water vapor pressure, J. Non-Cryst. Sol. **347** (2004) 211-219.

2.25 S. Ito, M. Tomozawa, Stress corrosion of silica glass, Comm. Am. Ceram. Soc. (1981), C160.

2.26 A. Agarwal, M. Tomozawa, Correlation of silica glass properties with the infrared spectra, Journal of Non-Crystalline Solids **209** (1997), 166–174.

2.27 Scholze, H., Glas, Springer Verlag, Berlin, 1988.

2.28 T. Zhu, J. Li, X. Lin, S. Yip, Stress-dependent molecular pathways of silica-water reaction, J. of Mechanics and Physics of solids, **53**(2005), 1597-1623.

3 Effect of stress-dependent diffusivity

3.1 Computation of diffusion profiles for temperatures <500°C

The diffusivity is a function of stress, commonly expressed by the hydrostatic stress component, σ_h, given by eq.(2.1.23). The diffusivity for the case of stress-enhanced diffusion is given by the following equation [3.1]

$$D = D_0 \exp\left[\sigma_h \frac{\Delta V_w}{RT}\right] \tag{3.1.1}$$

where D_0 denotes the value of the diffusivity in the absence of a stress. T is the absolute temperature in K; ΔV_w is the activation volume for stress-enhanced diffusion and R is the universal gas constant.

When an externally applied stress σ_{appl} and swelling stresses σ_{sw} are present simultaneously, the total hydrostatic stress is

$$\sigma_h = \sigma_{h,appl} + \sigma_{h,sw} \tag{3.1.2}$$

The swelling stress depends linearly on the water concentration, $\sigma_h \propto C$. The saturation value of $\sigma_{h,sw}$ for C=C_0 is in the following considerations denoted as $\sigma_{h,0}$.

Due to the swelling stresses, the diffusivity is a function of water concentration. Consequently, diffusion of water is now governed by the partial differential equation for the uniaxial diffusion

$$\frac{\partial C}{\partial t} = \frac{\partial}{\partial z}\left[D(C)\frac{\partial C}{\partial z}\right] \tag{3.1.3}$$

Equation (3.1.3) was numerically solved using the *Mathematica* procedure *NDSolve*

[3.2]. Figure 3.1a shows water profiles as a function of time with the actual water concentrations normalized on the value at the surface $C(0)$. The solid curves represent the solution in the absence of any stress effects on the diffusivity. With increasing time, the widths of the diffusion profiles increase. The bold dash-dotted curve in Fig. 3.1a is the limit case for saturation conditions $C(0) = C_0$ [3.3]:

$$C = C_0 \operatorname{erfc}\left(\frac{z}{2\sqrt{Dt}}\right) \tag{3.1.4}$$

In Fig. 3.1b the depth $z_{1/2}$ at which the water concentration decreased to the half of the surface value is plotted as a function of time. Whereas the diffusion zones in the absence of swelling stresses increase with time, they are clearly reduced with increasing time in the case of swelling included. This effect was already visible from Fig. 3.1a.

For reasons of simplicity, we introduce a normalized dimensionless time τ and normalized depth coordinate ζ, defined by

$$\tau = \frac{h^2}{D_0} t\,;\ \zeta = \frac{z}{\sqrt{D_0 t}} \tag{3.1.5}$$

Equations (1.2.5) and (1.2.6) then read

$$\frac{C(\zeta,\tau)}{C_0} = \operatorname{erfc}\left[\frac{\zeta}{2}\right] - \exp[\zeta\sqrt{\tau} + \tau]\operatorname{erfc}\left[\frac{\zeta}{2} + \sqrt{\tau}\right] \tag{3.1.6}$$

and

$$C(0,\,\tau)/C_0 = 1 - \exp[\tau]\operatorname{erfc}[\sqrt{\tau}], \tag{3.1.7}$$

In the absence of swelling, analytical solutions for the limit-cases τ=0 and $\tau\rightarrow\infty$ can be derived as was outlined in detail in [3.4, 3.5]. For the ratio $C(\zeta,\tau)/C(0,\tau)$ it results from eqs.(3.1.6) and (3.1.7)

$$\frac{C(\zeta,\tau)}{C(0,\tau)} = \frac{\operatorname{erfc}\left[\frac{\zeta}{2}\right] - \exp[\zeta\sqrt{\tau} + \tau]\operatorname{erfc}\left[\frac{\zeta}{2} + \sqrt{\tau}\right]}{1 - \exp[\tau]\operatorname{erfc}[\sqrt{\tau}\ ]} \tag{3.1.8}$$

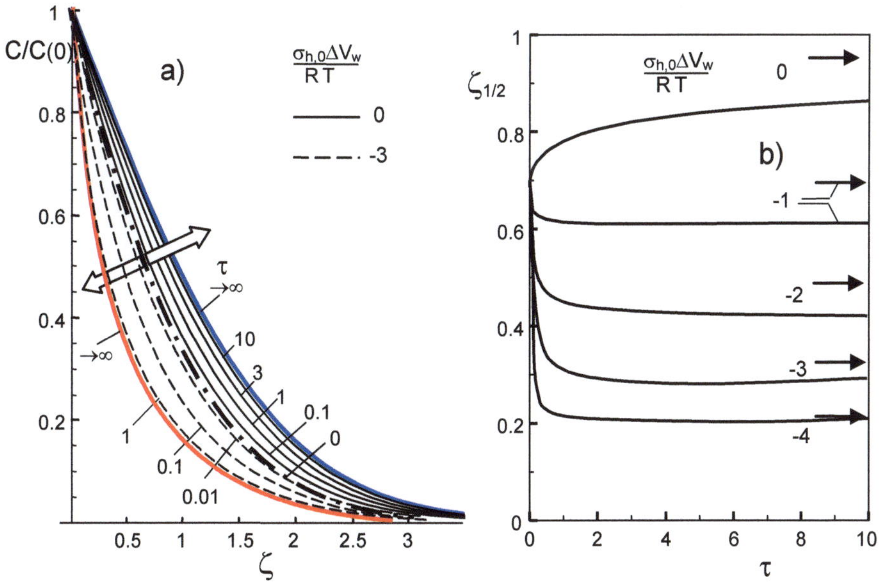

Fig. 3.1 a) Diffusion profiles as a function of time without and with consideration of swelling stress effect on diffusivity, b) diffusion depth $\zeta_{1/2}$ versus time, arrows indicate limit values for $\tau\to\infty$.

At very short times, we obtain by a series expansion with respect to τ that by setting $\tau\to 0$ reads

$$\frac{C(\zeta,0)}{C(0,0)} = \exp\left[-\frac{\zeta^2}{4}\right] - \frac{\sqrt{\pi}}{2}\zeta\,\mathrm{erfc}\left[\frac{\zeta}{2}\right] \tag{3.1.9}$$

At very long times $\tau\to\infty$, only the first term on the right-hand side of eq.(3.1.8) remains finite. Consequently, we obtain the well-known solution for constant surface concentration:

$$\frac{C(\zeta,\infty)}{C(0,\infty)} = \mathrm{erfc}\left[\frac{\zeta}{2}\right] \tag{3.1.10}$$

These limit cases are represented in Fig. 3.1a by the dash-dotted and the blue bold curve. The depths at which these limit distributions decrease to $C(\zeta,\tau)/C(0,\tau)=1/2$ are

$$\begin{aligned} \zeta_{1/2} &= 0.6995 \qquad \text{for} \quad \tau\to 0 \\ \zeta_{1/2} &= 0.9538 \qquad \text{for} \quad \tau\to\infty \end{aligned} \tag{3.1.11}$$

In presence of swelling, the solution for $\tau \to \infty$ is derived in [3.4, 3.5] and reads

$$\frac{C}{C_0} = \frac{RT}{\sigma_{h,0}\Delta V_w} \ln\left(1 - \left\{1 - \exp\left[\frac{\sigma_{h,0}\Delta V_w}{RT}\right]\right\} \operatorname{erfc}\left[\frac{\zeta}{2}\right]\right) \tag{3.1.12}$$

The red curve in Fig. 3.1a shows this solution for $\sigma_{h,0}\Delta V_w/RT$=-3. The depth $\zeta_{1/2}$ at which $C(\zeta)/C(0)$ decreases to $C(\zeta)/C(0)=C(\zeta)/C_0=1/2$, is

$$\zeta_{1/2} = 2\operatorname{erf}^{-1}\left[\infty, -\frac{1}{1+\exp[\frac{1}{2}\lambda]}\right] \tag{3.1.13}$$

where erf^{-1} is the inverse error function and the abbreviation λ stands for

$$\lambda = \frac{\sigma_{h,0}\Delta V_w}{RT}, \tag{3.1.14}$$

Numerical values for $\zeta_{1/2}$ are compiled in Table 1 in [3.5]. The limit values from (3.1.13) are introduced in Fig. 3.1b by the arrows.

3.2 Semi-analytical approach

The numerical effort for the *time-dependent* solution of the diffusion equation, eq.(3.1.3), in presence of swelling and stress-dependent diffusivity is hardly manageable. An approximate analytical solution can be obtained following the general procedure usual in perturbation theory. If we consider the diffusivity $D(C)$ as the disturbance parameter, perturbation theory suggests to solve the problem for the case of $D(0)=D_0$ and to insert the disturbance parameter into this solution. The disturbed solution is then obtained by introducing $D(C)$ instead of D_0 into the undisturbed solution. Since the analytical solution of the undisturbed problem, here denoted as the first-order solution $C^{(1)}$ is given by

$$\frac{C^{(1)}}{C_0} = \operatorname{erfc}\left[\frac{z}{2\sqrt{D_0 t}}\right] - \exp\left[\frac{h}{D_0}z + \frac{h^2}{D_0}t\right] \operatorname{erfc}\left[\frac{z}{2\sqrt{D_0 t}} + h\sqrt{\frac{t}{D_0}}\right] \tag{3.2.1}$$

the analytical second-order solution $C^{(2)}$ simply reads

$$\frac{C^{(2)}}{C_0} = \operatorname{erfc}\left[\frac{z}{2\sqrt{D(C^{(1)})t}}\right] - \\ - \exp\left[\frac{h}{D(C^{(1)})}z + \frac{h^2}{D(C^{(1)})}t\right]\operatorname{erfc}\left[\frac{z}{2\sqrt{D(C^{(1)})t}} + h\sqrt{\frac{t}{D(C^{(1)})}}\right] \tag{3.2.2}$$

with the diffusivity, eq.(3.1.1), taken at the local- and time-dependent concentration $C=C^{(1)}$.

A comparison of the perturbation solutions and the numerical results from eq.(3.1.3) is given in Fig. 3.2. Whereas for $\sigma_{h,0}\Delta V_w/RT=-2$ (a) the iteration solution $C^{(3)}$ is in complete agreement with the numerical solution, first small deviations are visible for $\sigma_{h,0}\Delta V_w/RT=-3$, Fig. 3.2b. It is therefore suggested to use the perturbation solution if the stress term is not too strong.

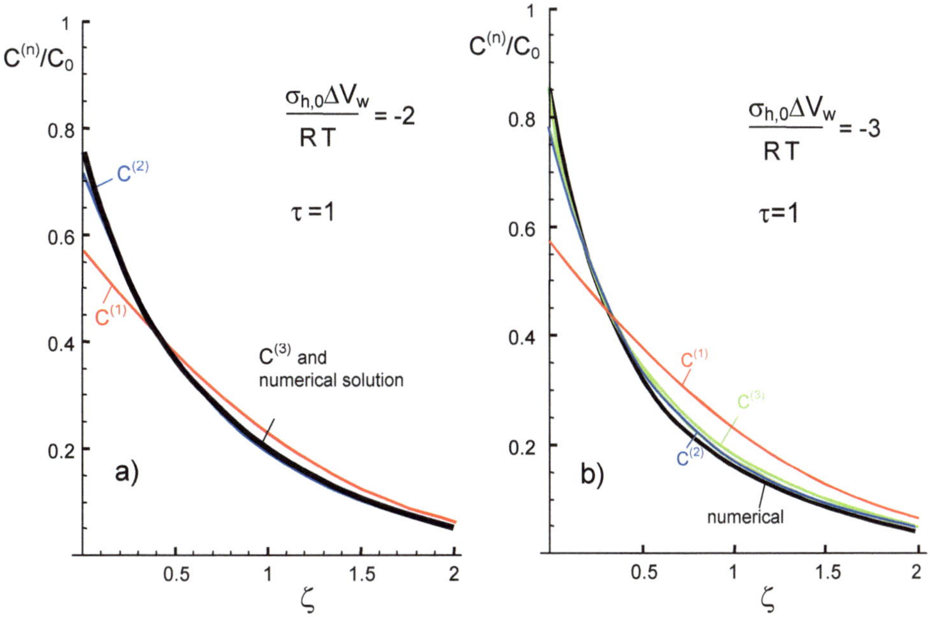

Fig. 3.2 a), b) Semi-analytical solutions $C^{(n)}$ compared with numerical results for $t\times h^2/D_0=1$.

Higher iteration solutions are possible by introducing $D(C(n))$ on the right-hand side, resulting in $C(n+1)$, etc.

$$\frac{C^{(n+1)}}{C_0} = \text{erfc}\left[\frac{z}{2\sqrt{D(C^{(n)})t}}\right] - \exp\left[\frac{h}{D(C^{(n)})}z + \frac{h^2}{D(C^{(n)})}t\right]\text{erfc}\left[\frac{z}{2\sqrt{D(C^{(n)})t}} + h\sqrt{\frac{t}{D(C^{(n)})}}\right] \tag{3.2.3}$$

Figure 3.3a shows the effect of the maximum swelling stress $\sigma_{h,0}$ on the shape of the concentration profiles. Figure 3.3b represents the time-dependent surface concentration normalized on the saturation concentration. The higher the compressive saturation swelling stress is the earlier is saturation reached.

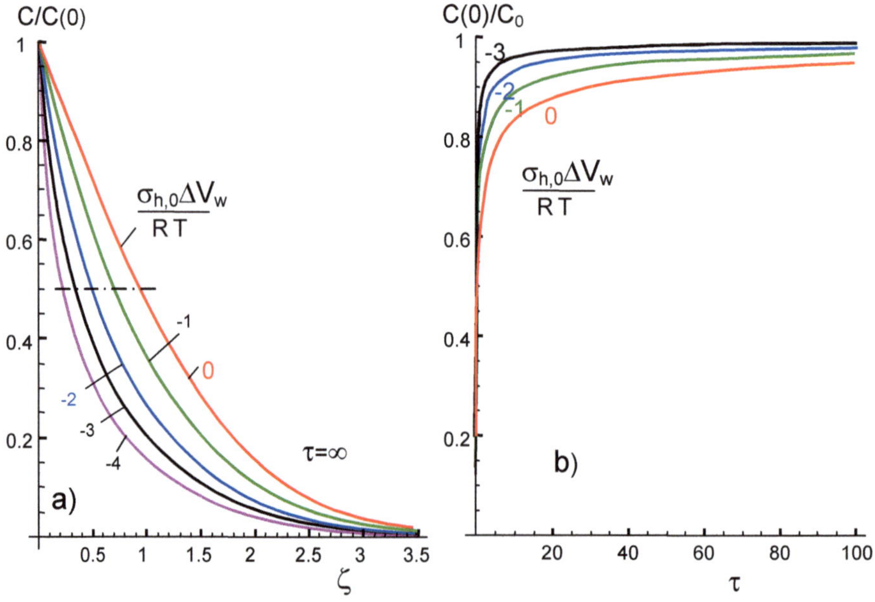

Fig. 3.3 a) Effect of the saturation stress $\sigma_{h,0}$, on the shape of the profiles for $\tau \rightarrow \infty$, b) surface concentrations vs. time.

3.3 Comparison with experimental results from literature

Profiles of time-dependent residual water content via stepwise surface removal were measured by Davis and Tomozawa [3.6] in water vapour of 355mm Hg pressure at 350°C. The results, Fig. 3.4, showed reduced diffusion depths with increasing soaking time. This behavior obviously agrees amazingly with the trend shown by the dashed curves in Fig. 3.1a. It can be interpreted as further evidence for the occurrence of swelling stresses.

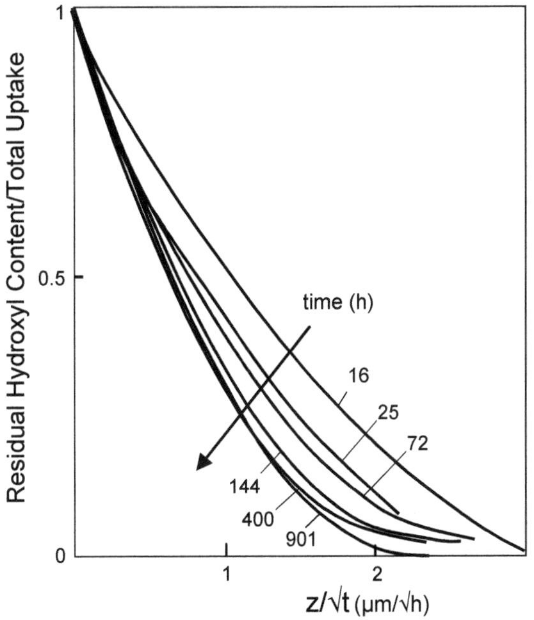

Fig. 3.4 Residual water content during stepwise surface removal Diffusion profiles by Davis and Tomozawa [3.6].

Effective diffusivity from diffusion profiles

Approximate methods for the determination of the diffusivity from measurements of water profile and water uptake were proposed by Tomozawa and co-workers [3.6, 3.7]. Basis of their evaluation is the use of an equation that is only valid for a *constant surface concentration*, although the concentrations clearly increased with time:

$$C = C(0)\,\mathrm{erfc}\left(\frac{z}{2\sqrt{D_{eff}\,t}}\right) \tag{3.3.1}$$

One possibility for the computation of an "effective diffusivity" is via the measured depth $z_{1/2}$ at which the concentration decreased to 50% of the surface concentration.

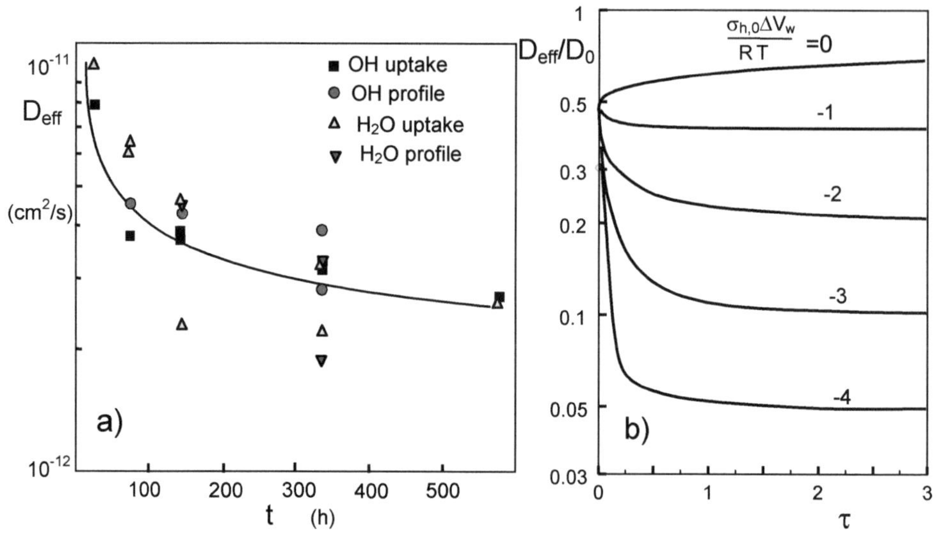

Fig. 3.5 a) Measurements of effective diffusivities by Oehler and Tomozawa [3.7] at 250°C and 39 bar water vapour pressure, b) effective diffusivity computed from the curves for the diffusion zone half-width $z_{1/2}$ of Fig. 3.1b.

3.4 Influence of externally applied stresses

Beside the swelling stresses also externally applied stresses as present for instance in bending tests must affect the diffusion profiles. The diffusivity in the presence of a hydrostatic swelling stress term σ_h and an externally applied hydrostatic stress $\sigma_{h,appl}$ is

$$D = D_0 \exp\left[(\sigma_{sw,h} + \sigma_{appl,h})\frac{\Delta V_w}{RT}\right] \tag{3.4.1}$$

For bending tests, the stresses are sufficiently constant within the range of the diffusion layers. In such tests, the upper and the lower surface of a bending bar show tensile and compressive stresses.

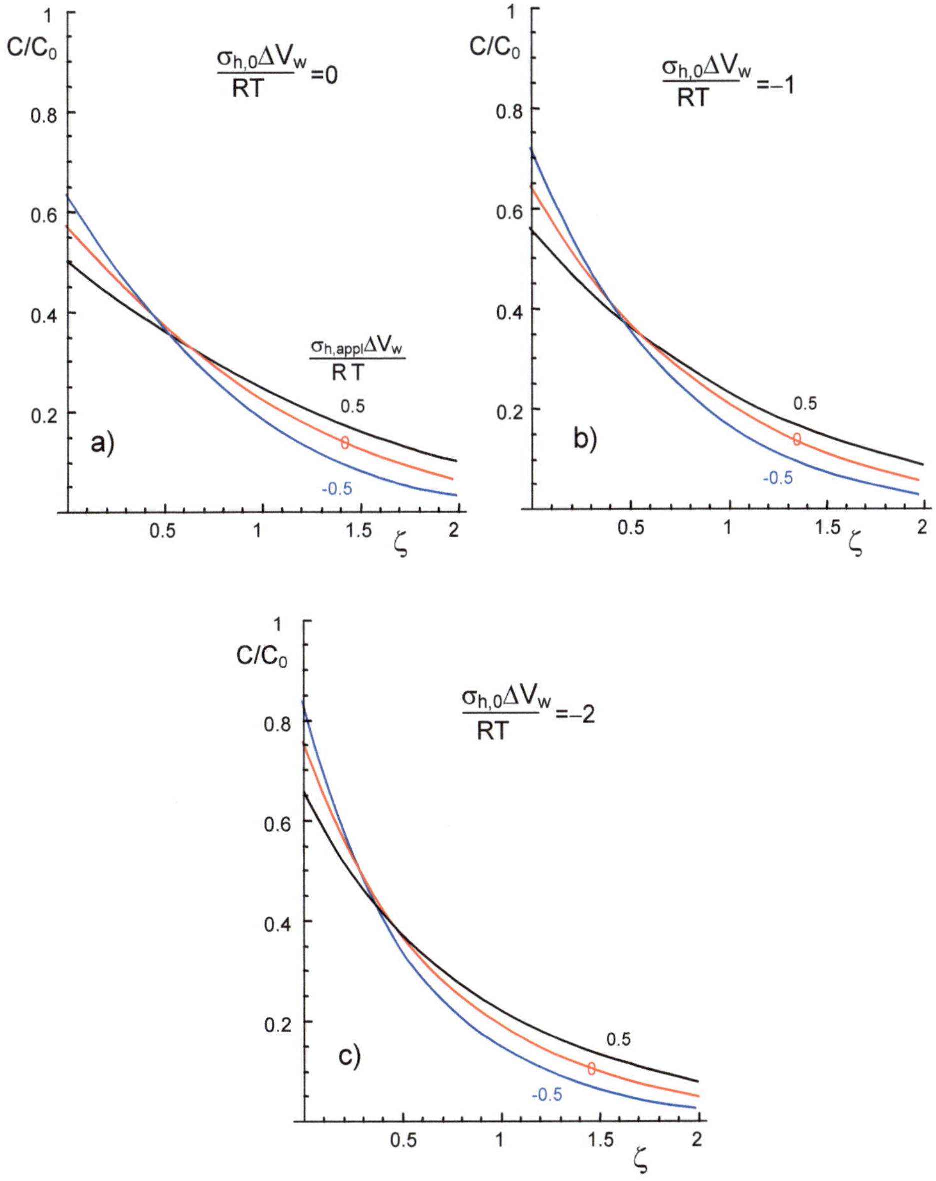

Fig. 3.6 Diffusion profiles under externally applied stresses; effect of swelling for soaking time τ=1.

Concentration profiles computed with eqs.(3.1.1) and (3.1.3) are plotted in Fig. 3.6 and Fig. 3.7 for varied hydrostatic swelling terms $\sigma_{h,0}\Delta V_w/RT$ and varied time. It is obvious that in all cases surface concentrations are smaller under externally applied tensile stresses than in the absence of such stresses. Compressive stresses cause an increase of hydroxyl water at the surface. The enlarged surface concentrations for compression decrease with a steeper profile than under tension. In deeper regions the concentrations under tension exceed those for the stress-free and the compression case.

Since the swelling stresses $\sigma_{h,0}$ depend on the saturation concentration which is proportional to the vapour pressure and since the activation volume ΔV_w in eq.(3.1.1) is not known sufficiently, the parameter $\sigma_{h,0}\Delta V_w/RT$ was tentatively varied in Fig. 3.6.

Comparison of Fig. 3.6a to Fig. 3.6c shows clearly that the reason of the difference in surface concentrations is not caused by the swelling stresses since the effect occurs even in the absence of swelling, (Fig. 3.6a). With increasing swelling the difference at the surface is only slightly increased. From Fig. 3.6a we therefore have to conclude that the effect of the externally applied stresses on surface concentrations is the increase of water with time, i.e. the parameter h in eqs.(1.2.4-1.2.6).

In Fig. 3.7 the time is varied for constant swelling $\sigma_{h,0}\Delta V_w/RT$=-2. The results indicate that for short times the concentration differences at the surfaces are larger than for longer times. For $\tau \rightarrow \infty$ the differences must of course disappear completely because then the water concentration reaches saturation independent on the applied stress.

The stress effect on the surface concentration can simply be summarized:
Under *tensile stresses* the diffusivity is increased and diffusion is enhanced. This must result in

- an <u>increased zone thickness</u> compared to tests without externally applied stresses.
- Due to the higher diffusivity under tensile stresses, the water diffusion into the bulk material is also increased. Since the water entrance from the environment into the surface is limited by the finite value of h, the concentration of water at the surface must decrease.

Under *compressive stresses* the diffusivity is reduced. This results in a reduced penetration depth and an increased surface concentration.

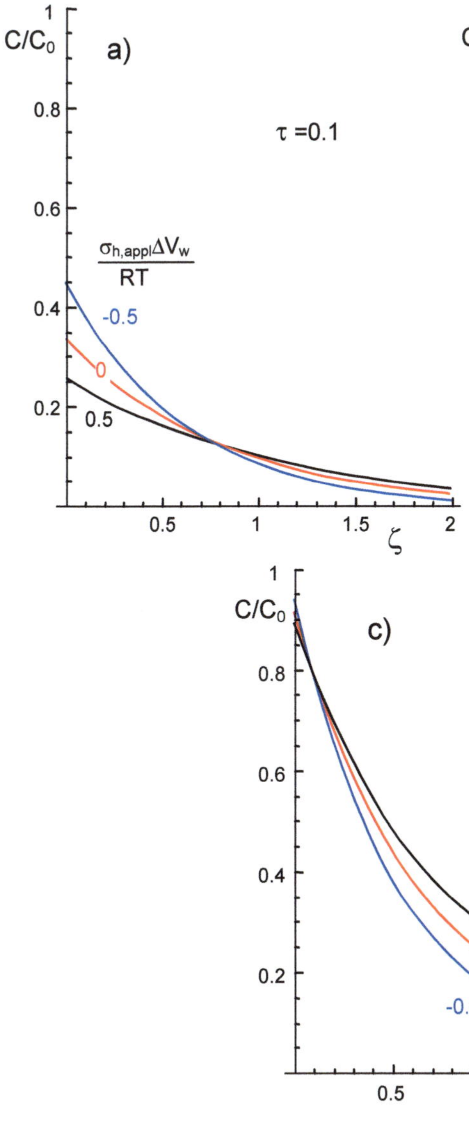

Fig. 3.7 Diffusion profiles under externally applied stresses; effect of normalized soaking time, τ, for $\sigma_{h,0}\Delta V_w/RT$=-2.

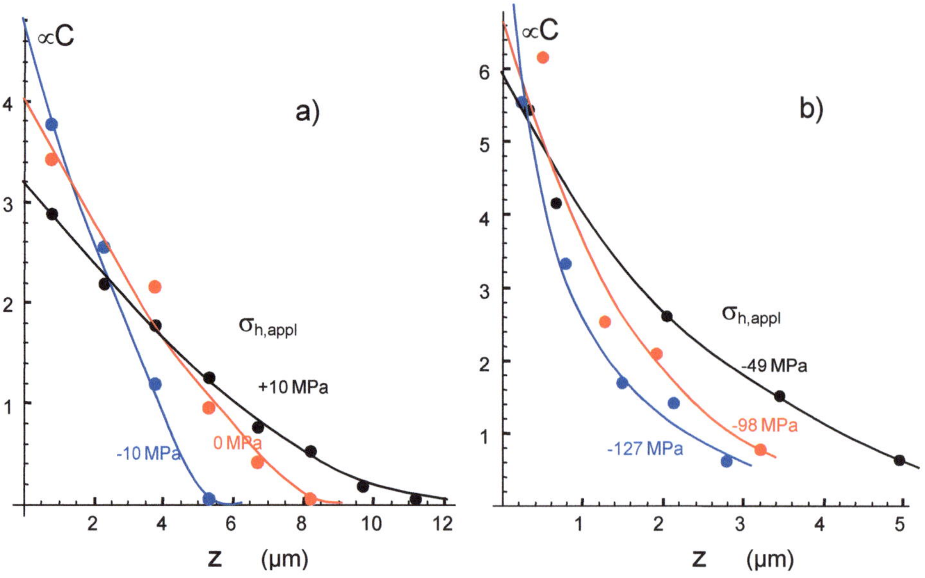

Fig. 3.8 Concentration profiles by Nogami and Tomozawa [3.8] as a function of the hydrostatic stress, a) bending bars for silica at 192°C (water vapour pressure: 12.3 bars), b) externally pressurized specimens at θ=350°C (water vapour pressure: 25 bars), C: in absorption coefficients (1/mm).

Figure 3.8a shows experimental results on bending bars by Nogami and Tomozawa [3.8] at 192°C carried out after 300h treatment in water atmosphere at 12 bars water vapour pressure. Measurements were carried out on the tensile and compression side under load and in an additional test also on unstressed bars.

In Fig. 3.8b similar results at 350°C are shown for 5h soaking (water atmosphere at 25 bars). In this case, Nogami and Tomozawa [3.8] applied hydrostatic pressures directly. The results of Fig. 3.8 completely confirm the computed curves in Fig. 3.6 and Fig. 3.7. The effective activation volume was experimentally determined in [3.8] as ΔV_{eff} =170 cm^3/mol for 192°C and 72 cm^3/mol at 350°C.

For the interpretation of the results in Fig. 3.5, the soaking temperature of 250°C is considered. Fig. 3.9a shows hydroxyl concentration S-data in mass units under water saturation pressure as could be computed from measurements by Zouine et al.[3.3] using the equilibrium equation from Wiederhorn et al. [3.9]. The straight line introduced in this plot reads

$$S_{\text{sat}} \cong 0.000265 \exp(0.0143\,\theta) \tag{3.4.2}$$

with the temperature θ in °C. An extrapolation to θ=250°C results in the saturation concentration of S=0.94 wt%. Under the assumption that even at moderate temperatures only hydroxyl generation contributes to the volume change, the corresponding hydrostatic swelling stress term at a free surface is

$$\sigma_{sw,h} = -\frac{2\kappa E}{9(1-\nu)} S = -18.7\text{ GPa} \times S \tag{3.4.3}$$

i.e. $\sigma_{h,0}$= -177 MPa. Consequently, it holds for 250°C:

$$\frac{\sigma_{h,0}\Delta V_w}{RT} = -0.0407\,\Delta V_w \tag{3.4.4}$$

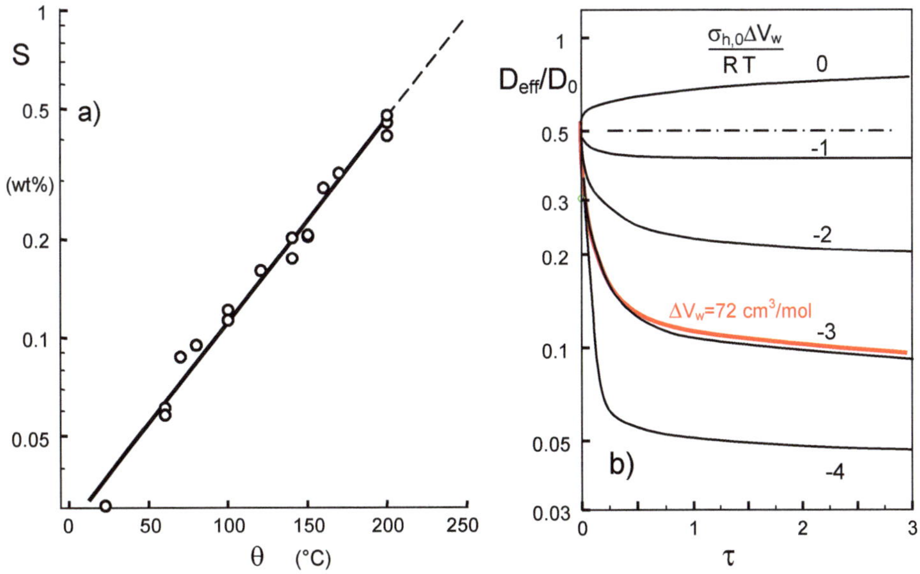

Fig. 3.9 a) Concentration of hydroxyl water at silica surfaces under saturation pressure by Zouine et al. [3.3], b) results of Fig. 3.5b together with the curve related to the effective volume ΔV=72 cm³/mol (red curve) as was derived by Nogami and Tomozawa [3.8].

In Fig. 3.9b the results of Fig. 3.5b are shown once more together with the parameter chosen for the results obtained by Nogami and Tomozawa [3.8], namely, ΔV_w =72 cm^3/mol (the value of ΔV_w=170 cm^3/mol seems to be extremely large). For the parameter (3.4.4) it results: $\sigma_{h,0}\Delta V_w/RT = -2.93$. The related D/D_0 dependency is shown in Fig. 3.9b by the red curve.

Finally, Fig. 3.10 illustrates the ratio D_{eff}/D_0 from Fig. 3.9 as a function of the swelling parameter $\sigma_{h,0}\Delta V_w/RT$ by the circles. The dash-dotted straight line represents eq.(3.1.1). The red arrows indicate the activation volumes suggested by Nogami and Tomozawa [3.8]. For strongly negative swelling parameters the integration procedure for the numerical solution of the differential equation (3.1.3) becomes rather unstable. A rough estimation for the very large value of 170 cm^3/mol needs an extrapolation of computed results. Such an extrapolation is tentatively shown by the dashed extension of the computed data curve.

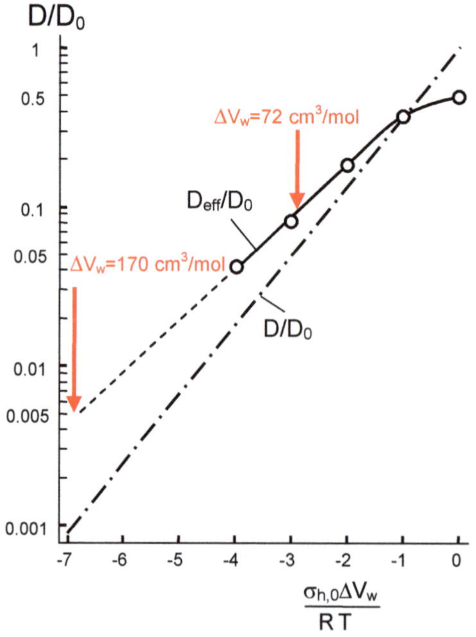

Fig. 3.10 Diffusivity ratio of D_{eff}/D_0 from Fig. 3.9b for normalized time τ=3 versus the swelling parameter $\sigma_{h,0}\Delta V_w/RT$ (symbols) compared with the diffusivity ratio D/D_0 according to eq.(3.1.1) (dash-dotted line); the dashed line is tentatively introduced for extrapolations.

A modified representation of the results in Fig. 3.6 and 3.7 was given in [3.10] as plotted in Fig. 3.11 for the lower temperature region <500°C, where simply $S \propto C$. In Fig. 3.11 the hydroxyl concentration at the surface, S, is normalized on the asymptotically reached value for disappearing external load, $S^{(0)}$, as is for instance fulfilled in the neutral axis of a bent bar. The hydroxyl content is under compression higher than under tension, Fig. 3.11a. If reaction (1.1.1) shows stress-enhancement as will be discussed in Section 5, the S-concentration in tension is by a factor $S \propto \exp[const \times \sigma / RT]$ increased and in compression decreased with the result given in Fig. 3.11b.

For long times when the water concentration approaches saturation, the S-content in tension exceeds that in compression due to the stress-enhanced equilibrium constant. The influence of stresses changes at the circle in Fig. 3.11b.

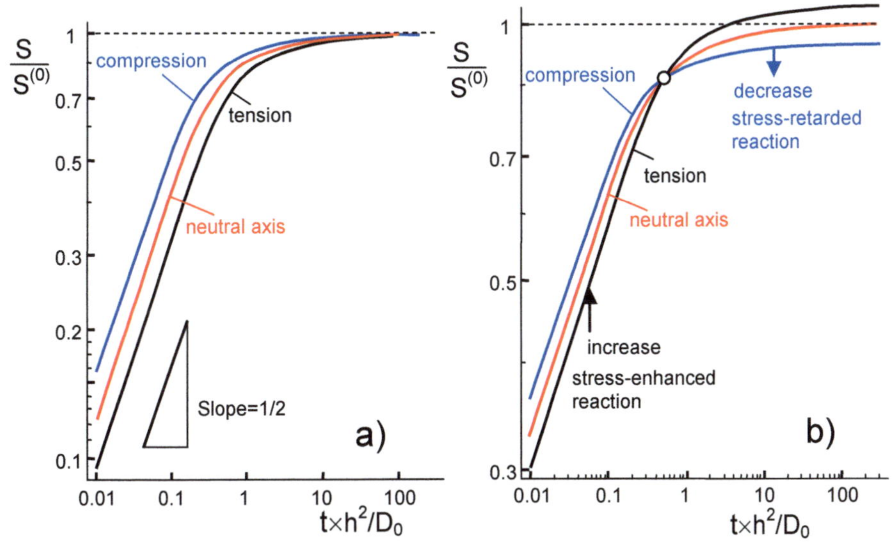

Fig. 3.11 a) Hydroxyl concentrations at a silica surface as a function of soaking time t based on the stress-affected diffusivity, exclusively; b) for stress-enhanced equilibrium of reaction (1.1.1). $S^{(0)}$ is the saturation concentration in the absence of external stresses. Diffusivity parameter $|\sigma_z|/3\times V_w/RT$=0.5, swelling parameter $\sigma_{sw,h,0} V_w/RT$=−3. Curves plotted for temperatures <500°C.

Hydrogen concentrations close to silica surfaces were reported by Agarwal et al. [3.11]. These authors used the nuclear reaction analysis (NRA) on water profiles in specimens that were hot-water soaked at 250°C under 355 Torr vapour pressure and a bending stress of 60 MPa. These results are represented in Fig. 3.12a. The water content under pressure (compressive side of a bent specimen) is always higher than under tension. This is in complete agreement with the data by Nogami and Tomozawa [3.8].

Since no surface concentrations are given in [3.11], we plotted in Fig. 3.12b the data at a depth of 0.12 μm below the surface versus soaking time. The slopes roughly fulfil the surface predictions from Fig. 3.11 for short times, namely $C/C_0 \propto \sqrt{t}$. No indication for saturation of the surface concentration is detectable.

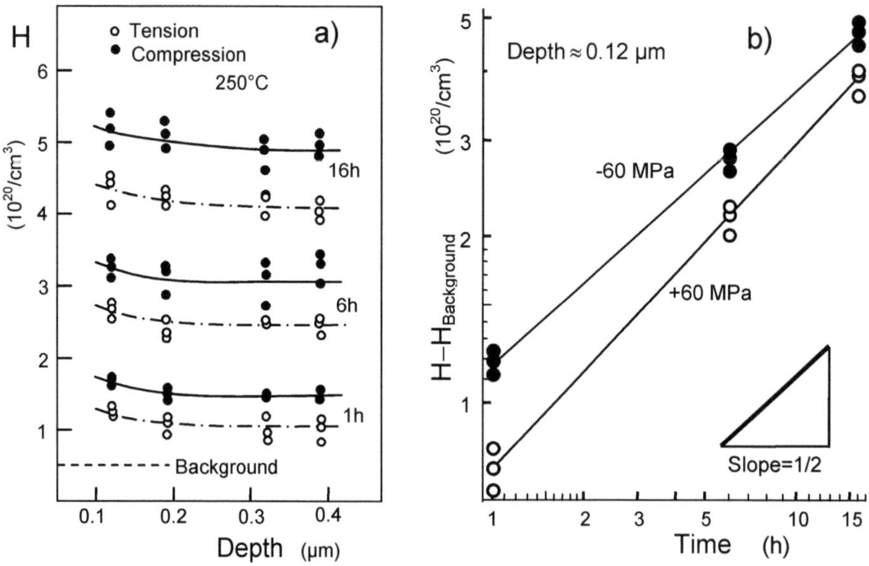

Fig. 3.12 a) Water concentration below the surface at 250°C and 355Torr vapour pressure under bending stresses of 60MPa, b) data at a depth of 0.12 μm below the surface plotted vs. soaking time.

Figure 3.13 shows the concentration of hydrogen atoms, H, in flexure specimens of silica glass, exposed to water vapour of 355 Torr at 650°C. The maximum bending stress in the specimens was ±45 MPa. Under stationary conditions, an increase of the concentration at the tension side and a decrease at the compressive surface were found.

These data confirm clearly the effect of higher water concentration under compression for short time and higher water concentration in tension for longer times. They are also in agreement with curves in Fig. 3.11a. The only difference is caused by taking the square-root for the ordinate since $S \propto C^{1/2}$ for temperatures >500°C.

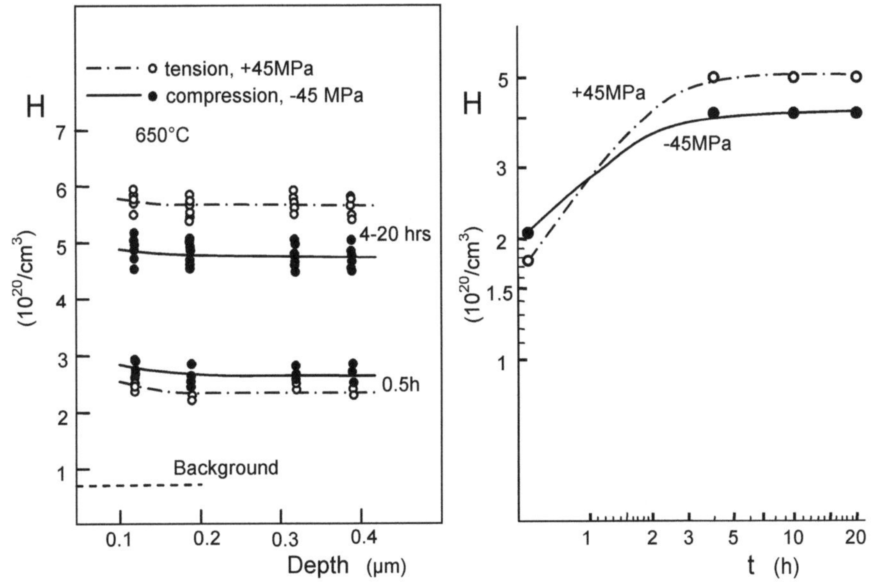

Fig. 3.13 Hydrogen concentration below the surface at 650°C and 355Torr vapour pressure under bending stresses of 45MPa.

Conclusions

There are several important findings from the computation of diffusion profiles for water vapour soaked silica. The computations show that:

- in the absence of swelling stresses the diffusion profile must tend for long time against the well-known description by the complementary error function;
- in presence of swelling stresses in the diffusion layer, the profiles must become steeper with increasing time;
- the effective diffusivity, D_{eff}, defined by the distance $z_{1/2} = \sqrt{D_{eff} t}$ at which the water concentration decreased to 50% of the surface value, must increase slightly if

no swelling stresses would occur. In presence of swelling, it should decrease strongly.

The agreement of computations and the experimental results by Davis and Tomozawa [3.6] and by Oehler and Tomozawa [3.7] clearly show that swelling of silica is present. From the computations, also the results by Nogami and Tomozawa [3.8] and Agarwal et al. [3.11] are easily understandable.

3.5 Diffusion in thin specimens

3.5.1 Computation of swelling stresses

The swelling stresses given by eqs.(2.1.22, 2.1.23) are valid for the case of thin swelling layers negligible compared with the specimen widths 2 *W*. Now let us consider thin specimens. In this case, the negative hydrostatic swelling stresses in the surface region where the diffusion zones develop must decrease after longer time. This is due to the fact, that all stresses must disappear when a constant water concentration is reached after long times. Then no restrictions on swelling are possible and the specimen can freely expand. The diffusivity trivially must tend to the value D_0 valid for σ_h=0.

The condition of mechanical equilibrium requires for the total stress distribution over the thickness 2 *W* that

$$\int_0^{2W} \sigma \, dz = 0 \tag{3.5.1}$$

This condition is fulfilled by the effective swelling stress σ_{eff}

$$\sigma_{eff}(z) = \sigma_{sw}(z) - \overline{\sigma}_{sw} = -\frac{1}{3}\frac{E}{1-\nu}(\varepsilon_v - \overline{\varepsilon}_v) \tag{3.5.2}$$

as can simply be validated by inserting (3.5.2) into (3.5.1).

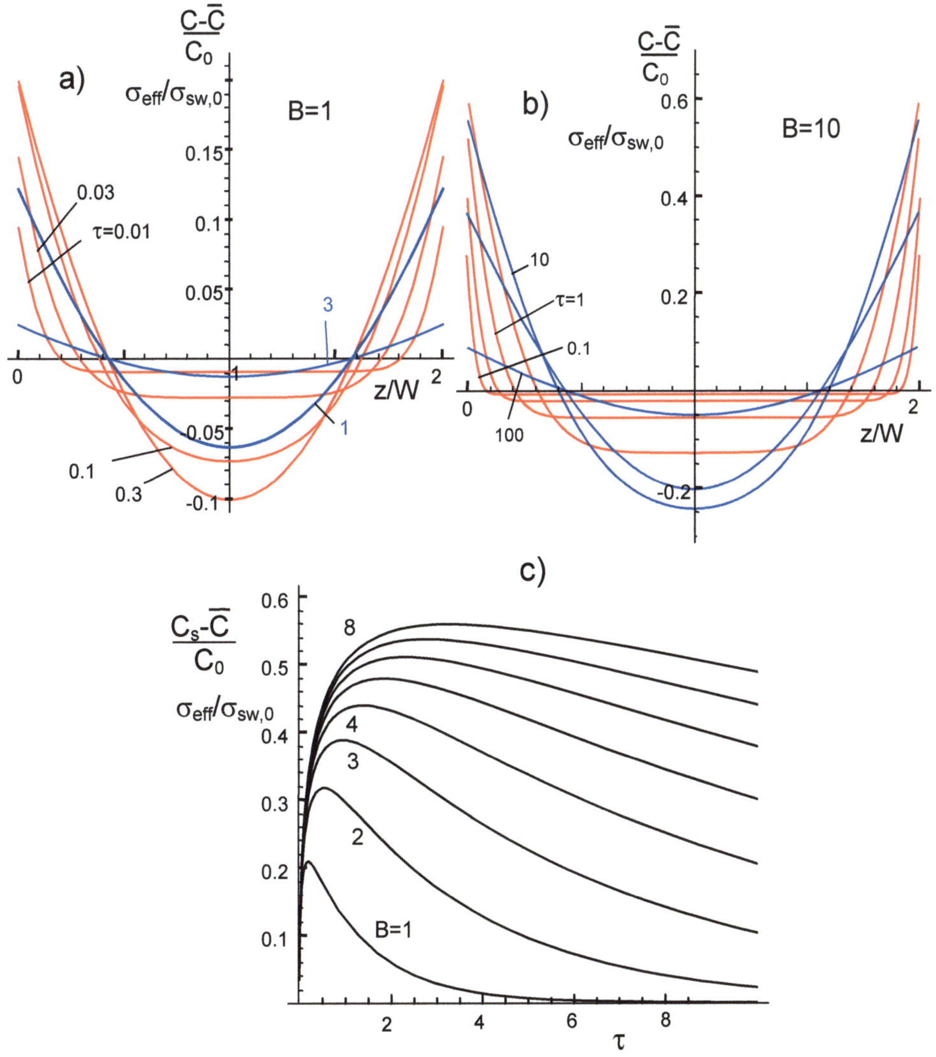

Fig. 3.14 a) Deviations of concentrations from their average over the width. Red curves: before maximum is reached, blue curves: after the maximum (the ordinates are proportional to the swelling stresses), b) concentrations for B=10 dependency of surface values with time, c) difference between surface and average concentration.

Figure 3.14 shows the influence of the time on the stress distributions (Figs. 3.14a, 3.14b) and the stresses at the surface (Fig. 3.14c). The stress values are scaled with the swelling stress $\sigma_{sw,0}$ that would occur at the surface of a semi-infinite body for $t\rightarrow\infty$. In case of the low-temperature reaction with equilibrium constant k_1 it holds for equi-biaxial swelling stress

$$\sigma_{sw,0} = -\frac{1}{3}\frac{E\kappa k_1 C_0}{1-\nu} \tag{3.5.3}$$

As a consequence of eqs.(3.5.2) and (3.5.3), it can be written with $\overline{C}$ defined by eq.(1.4.6)

$$\frac{\sigma_{eff}(z,t)}{\sigma_{sw,0}} = \frac{C(z,t)-\overline{C}(t)}{C_0} \tag{3.5.4}$$

For stress-enhanced swelling the equilibrium constant k_1 depends on the swelling stress. Then it must hold

$$\frac{\sigma_{eff}(z,t)}{\sigma_{sw,0}} = \frac{S(z,t)-\overline{S}(t)}{k_{1,0}C_0} \tag{3.5.5}$$

where $k_{1,0}$ is the equilibrium constant in the absence of any stress.

Figure 3.14a represents the deviations of the local water concentrations from the average value for a Biot number of B=1, i.e. the right hand side of eq.(3.5.4). The red curves show the results for times before the maximum surface value and the blue curves illustrate the distributions after the maximum. Figure 3.14c gives the surface values as a function of normalized time τ for different B.

3.5.2 Expectations from first-order analysis

Swelling of silica has several consequences on material behaviour. Two effects on diffusion behaviour may be addressed in this section. Since the analytical solution by the eqs.(1.4.1a) to (1.4.3a) does not include stress effects, these equations allow only to perform first-order estimations.

In [3.10] we could show that the surface water concentration increases at the surface under compressive stresses and decreases under tensile stresses. These findings are a simple consequence of the *limited mass-transfer* coefficient [3.10] or *slow-reaction* boundary condition [3.12] at the silica surface.

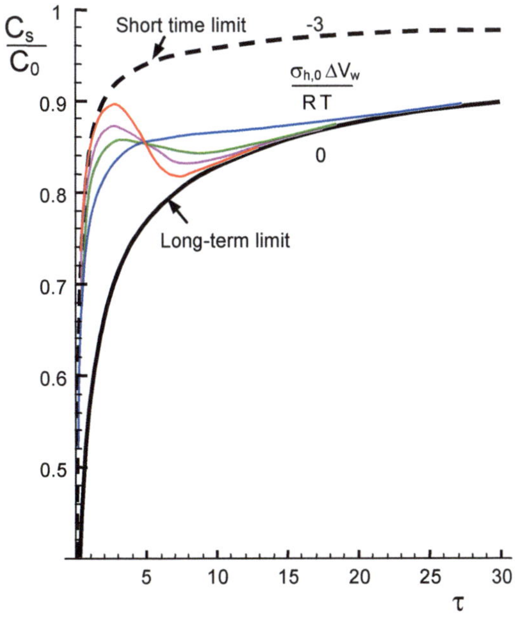

Fig. 3.15 Effect of diffusion boundary conditions at the silica surface and stress-affected diffusivity, eq.(3.4.1), on the normalized surface concentration of water, $C_s(\tau)/C_0$. Dashed curve: Limit case for the half-space under swelling at the surface; Solid curve: Limit case for diffusion in the absence of swelling stresses; Coloured curves: Possible limit-case interpolations.

Two limit cases may be considered:

(1) For very short times, t, the thickness of the water diffusion zone, $b=(D\times t)^{1/2}$ is small compared to the specimen thickness, $b<<W$. Consequently, diffusion behaves like in a semi-infinite body. This limit case was outlined in [3.10]. The surface concentration C_s for $\sigma_{h,0}\Delta V_w/RT$=-3 is shown in Fig. 3.15 as the dashed curve.

(2) For very long times, the specimen is completely soaked by water, resulting in the concentration $C(z, t)\rightarrow C_0$. Any swelling stress must disappear in the specimen as a consequence of mechanical equilibrium, eqs.(3.5.2, 3.5.4). The solution for the half-

space, computed with a disappearing stress-enhancement of $\sigma_{h,0}\Delta V_w/RT=0$, is represented in Fig. 3.15 by the solid limit curve.

The real diffusion behaviour must start on the dashed curve for short times and tend to the solid curve for long times. This is schematically indicated by the coloured thin curves. Depending on the material parameters this change can occur very slowly (blue curve) or more abruptly (red curve).

From Fig. 3.15 it can be expected that for long times for which the layer thickness b and the specimen width W become comparable, a maximum in the surface concentration $C(t)$ occurs.

3.5.3 Numerical results for stress-enhanced diffusivity

Equation (3.1.3) was numerically solved using the *Mathematica* procedure *NDSolve* [3.2]. Two boundary conditions had to be satisfied simultaneously, namely

$$\frac{dC}{dz} = -\frac{h}{D}(C_0 - C) \quad \text{at } z=0, \tag{3.5.6}$$

$$\frac{dC}{dz} = +\frac{h}{D}(C_0 - C) \quad \text{at } z=2W, \tag{3.5.7}$$

In the following considerations, it is assumed that the coefficient h might be independent of stress. The swelling stress related to a certain concentration C is

$$\sigma_{h,sw} = \sigma_{h,0}\frac{C}{C_0} \quad , \quad \sigma_{h,0} = \sigma_{h,sw}(C_0) \tag{3.5.8}$$

where the quantity $\sigma_{h,0}$ is the hydrostatic swelling stress value reached for $C=C_0$. Figure 3.16 shows the surface values of the effective swelling stresses defined by eq.(3.5.2) as a function of time τ, Biot number B and hydrostatic swelling stresses $\sigma_{h,0}\Delta V_w/RT=-3$ [3.13]. The upper limit curve with $B\rightarrow\infty$ represents the case of the half-space, since $B\propto W$. In this case, any diffusion layer is small compared to the specimen width and an unloading effect with increasing layer thickness is impossible.

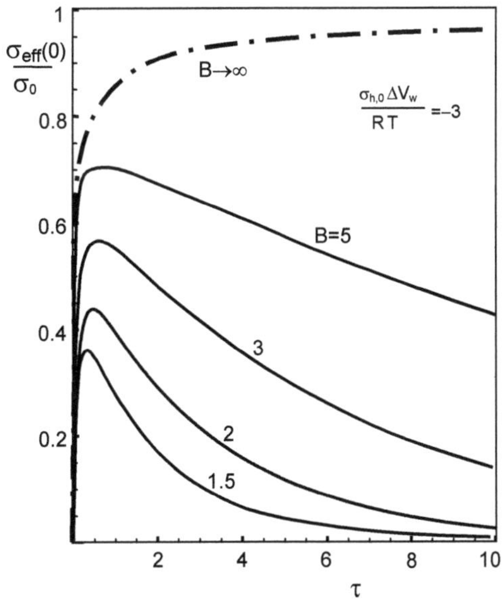

Fig. 3.16 Effective swelling stresses at the surface.

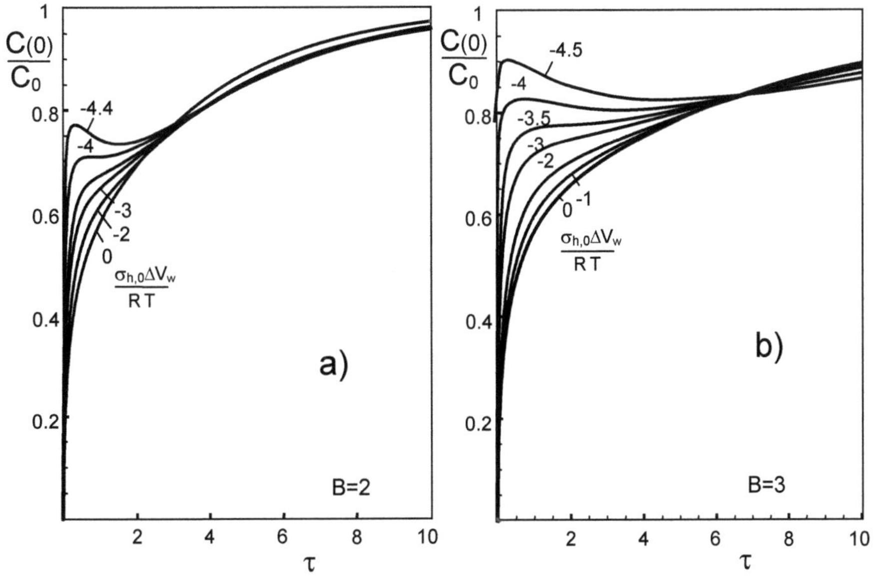

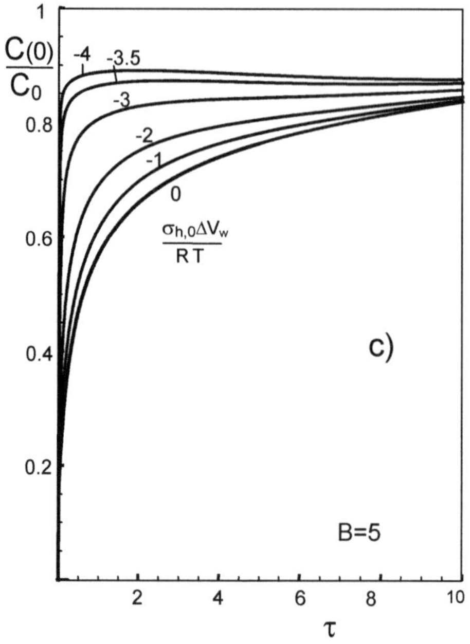

Fig. 3.17 Water concentration at the surface (effect of swelling stress, Biot number B, and time τ).

Figure 3.17 represents the surface concentration as a function of time τ and Biot number B. An interesting effect is visible for stronger swelling as illustrated in Figs. 3.17a-3.17c for several Biot numbers.

These figures show a relative concentration maximum, followed by a temporary decrease of $C(0)$. For large times the concentration C_0 is asymptotically approached.

It should be mentioned that for stronger swelling, $|\sigma_{h,0}\Delta V_w/RT|>4.5$, the *Mathematica* Routine *NDSolve* did no longer converge.

References

3.1 P.G. Shewman, Diffusion in Solids, McGraw-Hill, New York, 1963.

3.2 *Mathematica*, Wolfram Research, Champaign, USA.

3.3 A. Zouine, O. Dersch, G. Walter and F. Rauch, "Diffusivity and solubility of water in silica glass in the temperature range 23-200°C," *Phys. Chem. Glass: Eur. J. Glass Sci and Tech. Pt. B*, **48** [2] 85-91 (2007).

3.4 T. Fett, G. Schell, Analytical solutions of the diffusion differential equation, Scientific Working Papers **67**, 2017, ISSN: 2194-1629, Karlsruhe, KIT.

3.5 Schell, K.G., Fett, T., Bucharsky, E.C., Diffusion equation under swelling stresses, SN Springer Nature Switzerland (2019), 1:1300, online: |https://doi.org/10.1007/s42452-019-1343-1.

3.6 Davis, K.M., Tomozawa, M., Water diffusion into silica glass: structural changes in silica glass and their effect on water solubility and diffusivity, J. Non-Cryst. Sol. **185**(1995), 203-220.

3.7 Oehler, A., Tomozawa, M., Water diffusion into silica glass at a low temperature under high water vapor pressure, J. Non-Cryst. Sol. **347**(2004) 211-219.

3.8 M. Nogami, M. Tomozawa, Diffusion of water in high silica glasses at low temperature, Phys. and Chem. of Glasses **25**(1984), 82-85.

3.9 S. M. Wiederhorn, F. Yi, D. LaVan, T. Fett, M.J. Hoffmann, Volume Expansion caused by Water Penetration into Silica Glass, J. Am. Ceram. Soc. **98**(2015), 78-87.

3.10 S. M. Wiederhorn, G. Rizzi, G. Schell, S. Wagner, M. J. Hoffmann, T. Fett, Diffusion of water in silica: Influence of moderate stresses, J. Am. Ceram. Soc. **101**(2015), 1180-1190.

3.11 A. Agarwal, M. Tomozawa, W. A. Lanford, Effect of stress on water diffusion in silica glass at various temperatures, J. Non-Cryst. **167**(1994), 139-148.

3.12 R.H. Doremus, Diffusion of Reactive Molecules in Solids and Melts, Wiley, 2002, New York.

3.13 T. Fett, G. Schell, Water diffusion in thin silica sheets, Effect of swelling stress on diffusi-vity, Scientific Working Papers **72**, 2017, ISSN: 2194-1629, Karlsruhe, KIT.

4 Surface concentrations and swelling stresses

4.1 Hydroxyl concentration under water saturation pressure

4.1.1 Results from literature

Hydroxyl concentrations at silica surfaces below 500°C are available from investigations by Tomozawa and co-workers [4.1,4.2] for water vapour pressure of 355 Torr and for water saturation pressure by Oehler and Tomozawa [4.3] and Zouine et al. [4.4].

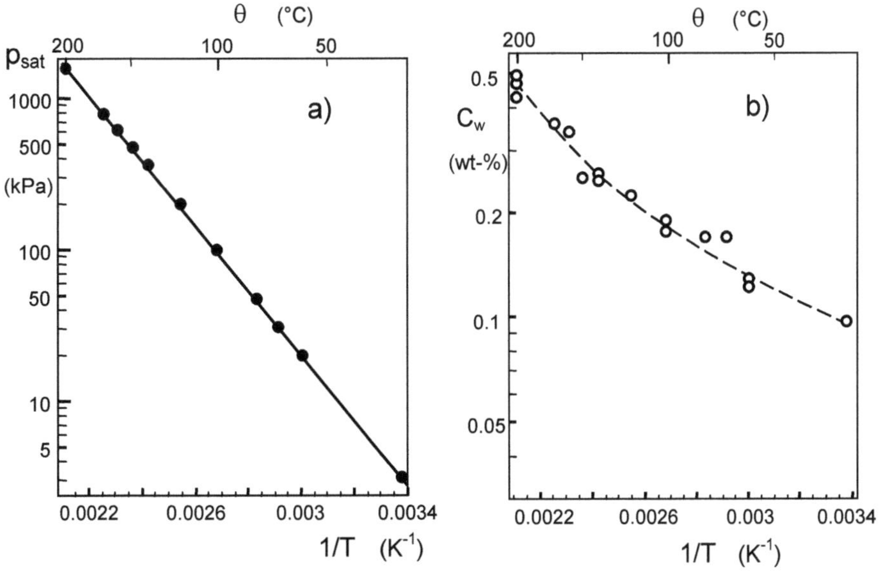

Fig. 4.1 Results by Zouine et al. [4.4], a) saturation pressure with symbols indicating the test conditions, b) solubility of water at silica surfaces under saturation pressure.

The results from Zouine et al. [4.4] are shown in Fig. 4.1. The saturation pressures *p*, under which the tests were carried out, are represented in Fig. 4.1a. For these measurements the *nuclear reaction analysis* (NRA) was applied that subsumes molecular and hydroxyl water. Whereas in [4.1-4.3] the *S*-concentration was measured via the IR-method, the data by Zouine et al. [4.4] represent the total water content C_w at the surface. The results are shown in Fig. 4.1b. From these measurements, also the water species *S* and *C* can be obtained.

4.1.2 Computation of molecular and hydroxyl water concentrations

In *molar units*, the total water concentration is given by

$$C_w = C + \tfrac{1}{2}S = C(1 + \tfrac{1}{2}k) \tag{4.1.1}$$

where the quantity *k* is the equilibrium constant describing the ratio of $k=S/C$. The experimental results on equilibrium ratios from literature were expressed in [4.5] for the temperature range of 90°C≤T≤350°C by the empirical relation

$$k = \frac{S}{C} = A\exp\left(-\frac{Q}{RT}\right) \tag{4.1.2}$$

(A=32.3 and Q=10.75 kJ/mol). Equations (4.1.1) and (4.1.2) result in

$$C = \frac{C_w}{1 + \tfrac{1}{2}k}, \tag{4.1.3}$$

$$S = \frac{C_w}{\left(\dfrac{1}{2} + \dfrac{1}{k}\right)} \tag{4.1.4}$$

and in *mass units* considering the different mole masses of water and hydroxyl

$$S = \frac{17}{18}\frac{C_w}{\left(\dfrac{1}{2} + \dfrac{1}{k}\right)} \tag{4.1.4a}$$

Dividing C_w, C, and S by the density of silica yields the water species in parts per mass of glass, mostly in ppm or wt-%. For the temperature range of 20°C≤θ≤200°C, the straight-line dependence of saturation pressure, Fig. 4.1a, can be approximated as

$$p_{sat} \cong B_1 \exp\left(-\frac{B_2}{RT}\right) \tag{4.1.5}$$

(B_1=5.79×10^7 kPa, B_2=41.28 kJ/mol). Under the common assumption that the concentration of molecular water species at the surface is proportional to the pressure of the water vapour, $C \propto p$, the combination with the two $1/T$-depenencies for k and p makes clear that the total water concentration at the surface, eq.(4.1.1) not necessarily shows a straight line in an Arrhenius plot. This behaviour is visible from the dashed curve in Fig. 4.1b. The concentrations of molecular and hydroxyl water were computed from the measured total water concentration by using eq.(4.1.3) and (4.1.4a) with the result plotted in Fig. 4.2 versus $1/T$. The dashed curves are guidelines for the eyes.

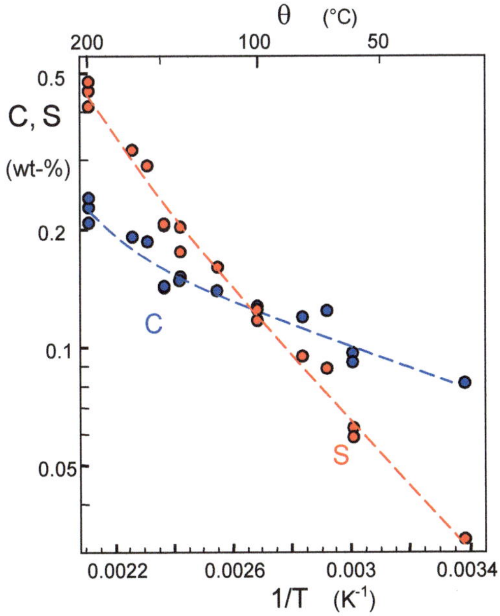

Fig. 4.2 Surface concentration of the molecular (blue) and hydroxyl (red) water species obtained via eqs.(4.1.3), (4.1.4a).

4.1.3 Semi-logarithmic description

Figure 4.3a shows the total water concentration in a semi-logarithmic plot with linear temperature scaling. The measurements can approximately be described by the simple expression

$$C_w = 0.000780\exp(0.00868\,\theta) \quad \text{for } 23°\text{C} \le \theta \le 200°\text{C} \tag{4.1.6}$$

with the temperature θ in °C.

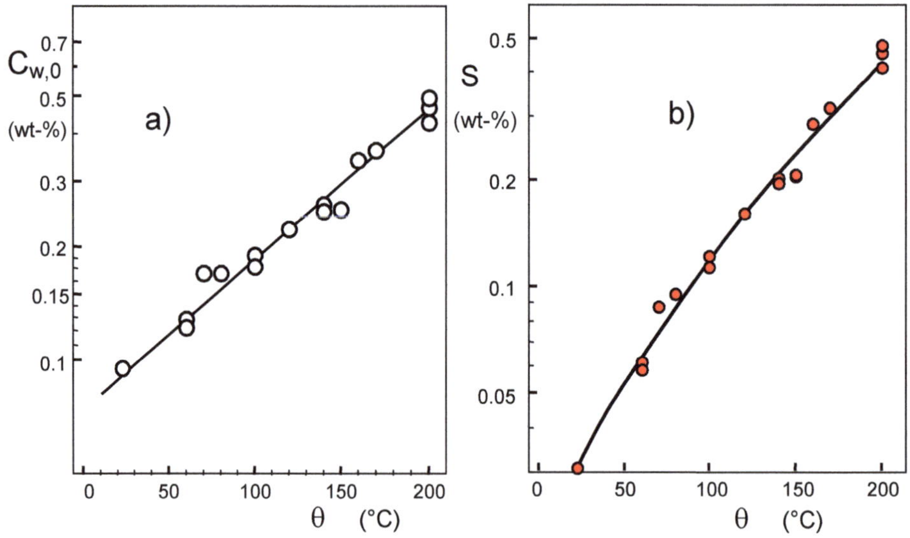

Fig. 4.3 a) Total water from Fig. 4.1b and hydroxyl data of Fig. 4.2 plotted with linear abscissa scaling, b) hydroxyl water concentration, curve: eq.(4.1.7) obtained by combining eqs.(4.1.6) and (4.1.2).

Figure 4.3b illustrates the *S*-data in a similar plot with a slightly curved averaging curve given by

$$S = \tfrac{17}{18}\frac{0.000780\exp(0.00868\,\theta)}{\frac{1}{2}+\frac{1}{A}\exp(Q/RT)} \tag{4.1.7}$$

with the parameters *A* and *Q* in eq.(4.1.2). For simple computations in the reduced temperature range of 70°C≤θ≤200°C, tolerating a few percentage deviations from the curve in Fig. 4.3b, a straight-line representation is suggested that reads

$$S_{\text{sat}} \approx 0.000265 \exp(0.0143\,\theta) \quad \text{for } 70°\text{C} \le \theta \le 200°\text{C} \tag{4.1.7a}$$

This approximation is introduced in Fig. 4.4a as the red line. For extrapolations, eq.(4.1.7a) should not be used. Already at θ=250°C, the hydroxyl concentration is overestimated by about 30%.

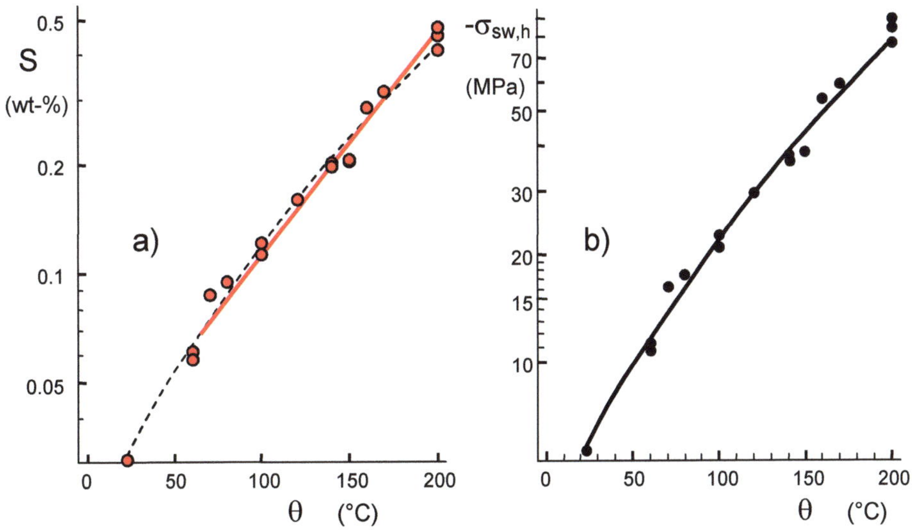

Fig. 4.4 a) Hydroxyl data of Fig. 4.3b together with the simple straight-line approximation by eq.(4.1.7a), b) Hydrostatic swelling stresses under saturation water pressure.

The swelling stresses at the glass surfaces are given by eqs.(2.1.11) and (2.1.22) as

$$\sigma_{sw,z} = \sigma_{sw,y} = -\kappa \frac{E}{3(1-\nu)} S \cong -28 \text{ GPa} \times S \tag{4.1.8}$$

Introducing the hydroxyl concentrations from Fig. 4.2 results in the swelling stresses plotted in Fig. 4.4b. For the computation of the swelling stresses, the necessary hydroxyl concentration can be determined via eq.(4.1.4a) with the equilibrium constant according to eq.(4.1.2). Introducing the hydroxyl concentrations from eq.(4.1.7) results in the swelling stresses

$$\sigma_{sw,h} = -\frac{13.75\,\text{MPa}\,\exp(0.00868\,\theta)}{\frac{1}{2} + \frac{1}{A}\exp(Q/RT)} \tag{4.1.9}$$

4.2 Hydroxyl concentration under constant pressure

In order to eliminate the effect of increasing surface concentrations with increasing saturation pressure, the concentrations were normalized on the pressure of 355 Torr (see also [4.4]). In contrast to Fig. 4.2, the surface concentrations decrease with increasing temperature. This makes clear that the increasing curves of Fig. 4.1b and Fig. 4.2 reflect strongly the increasing saturation pressure.

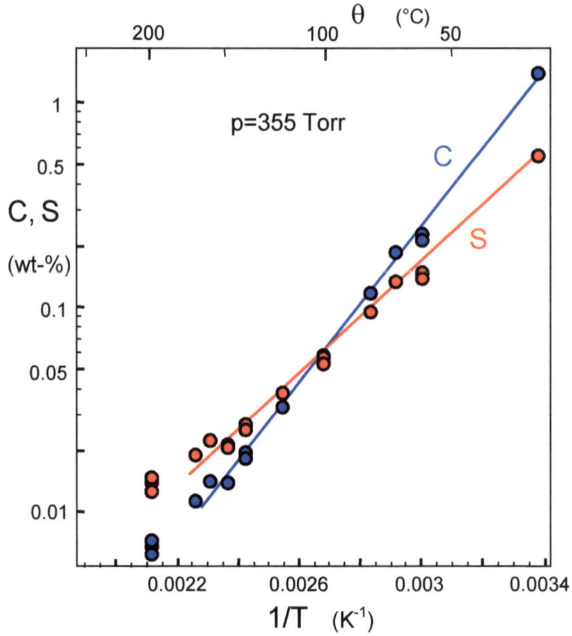

Fig. 4.5 Concentrations of molecular and hydroxyl water from Fig. 4.2, normalized on a constant water vapour pressure of $p = 355$ Torr.

The data in Fig. 4.5, describing concentrations of molecular water and hydroxyl in the absence of externally applied stresses, can be represented by straight lines for $1/T > 0.0022/\mathrm{K}$:

$$C_{355\,\mathrm{Torr}} \cong \exp\left[-18.94 + \frac{35.84\,\mathrm{kJ/mol}}{RT}\right] \tag{4.2.1}$$

and

$$S(0)_{355\,\mathrm{Torr}} \cong \exp\left[-15.56+\frac{25.09\,\mathrm{kJ/mol}}{RT}\right] \tag{4.2.2}$$

References

4.1 Wakabayashi, H., Tomozawa, M., Diffusion of water into silica glass at low temperature, J. Am. Ceram. Soc. **72**(1989), 1850-55.

4.2 Davis, K.M., Tomozawa, M., Water diffusion into silica glass: structural changes in silica glass and their effect on water solubility and diffusivity, J. Non-Cryst. Sol. **185** (1995), 203-220.

4.3 Oehler, A., Tomozawa, M., Water diffusion into silica glass at a low temperature under high water vapor pressure, J. Non-Cryst. Sol. **347**(2004) 211-219.

4.4 A. Zouine, O. Dersch, G. Walter and F. Rauch, "Diffusivity and solubility of water in silica glass in the temperature range 23-200°C," *Phys. Chem. Glass: Eur. J. Glass Sci and Tech. Pt. B*, **48** [2] 85-91 (2007).

4.5 S. M. Wiederhorn, F. Yi, D. LaVan, T. Fett, M.J. Hoffmann, Volume Expansion caused by Water Penetration into Silica Glass, J. Am. Ceram. Soc. **98** (2015), 78-87.

5 Stress-enhanced equilibrium

5.1 Thermodynamic description

At temperatures $T > 500°C$, the equilibrium constant of the reaction

$$\equiv Si\text{-}O\text{-}Si\equiv + H_2O \leftrightarrow \equiv SiOH + HOSi\equiv \quad (5.1.1)$$

is

$$k_2 = \frac{S^2}{C}. \quad (5.1.2)$$

According to Le Chatelier [5.1], the equation governing the equilibrium constant is

$$\frac{\partial \ln k}{\partial p} = -\frac{\Delta \overline{V}}{RT}, \quad \Delta \overline{V} = 2\overline{V}_S - \overline{V}_C. \quad (5.1.3)$$

where p is pressure, $\Delta \overline{V}$ is the change in the partial molar volume of the reaction, $\overline{V}_S$ and $\overline{V}_C$ the partial molar volumes of hydroxyl and molecular water, respectively. R is the universal gas constant and T is the temperature in °K. At high temperatures, some of the water will react with the silica network according to eq.(5.1.1) to form hydroxyl groups, $S = [\equiv SiOH]$, in the silica network. The concentration of the unreacted water is $C = [H_2O]$. We believe that the molecular water in the glass is located in holes of molecular size of the silica glass network [5.2], where the water does not contribute to the volume of the glass. As a consequence, molecular water cannot contribute to the stress effect. Thus, the concentration of molecular water in the glass depends only on the vapour pressure in the external environment, but not on the stress in the glass.

This same conclusion was reached earlier by Agarwal *et al.* [5.3] using slightly different reasoning. They noted that noble gases such as helium diffuse through silica glass with little sensitivity to the applied stress [5.4]. Water molecules are believed to diffuse through silica glass in much the same way as the noble gases. Hence, the diffusion of water molecules is also expected to show little dependence on applied stress. Following this reasoning, all of the sensitivity of the hydrogen concentration to the applied stress in the experiments by Argawal *et al.* [5.3] resides in the hydroxyl concentration.

This is equivalent to $\overline{V}_C = 0$ in eq.(5.1.3). Consequently, molecular water cannot contribute to the stress effect as had been shown by deformations on water-soaked disks [5.5]. Thus, the concentration of molecular water in the glass depends only on the vapour pressure in the external environment, but not on the stress in the glass. By replacing the hydrostatic pressure p by the hydrostatic stress σ_h in a solid

$$\sigma_h = \tfrac{1}{3}(\sigma_x + \sigma_y + \sigma_z) \tag{5.1.4}$$

we obtain with the hydroxyl concentration S_0 for σ_h=0

$$S = S_0 \exp\left(\frac{\sigma_h \overline{V}_S}{RT}\right) \quad , \quad \sigma_h = -p \tag{5.1.5}$$

At temperatures T< 450°C, hydroxyls are immobile. Only those hydroxyls can react in the reverse reaction step, which are directly neighboring. Consequently, the reaction is of first order. In this case the equilibrium constant for the reaction is

$$k = \frac{S}{C}. \tag{5.1.6}$$

The result for the hydroxyl concentration is therefore of the same form as given by eq.(5.1.5) possibly with a different reaction volume $\overline{V}$ in the different temperature regions of <500°C and >500°C.

5.2 Experimental evidence for stress-enhancement

5.2.1 Water concentration at crack surfaces of subcritically grown cracks

An interesting experiment was performed by Tomozawa et al. [5.6]. Fracture surfaces were formed by the passage of a crack through Double Cleavage Drilled Compression specimens (DCDC) [5.7, 5.8], while the specimen was exposed to water. The experiments demonstrated that the surface concentration of water was significantly higher near the fracture surface than would be the case for silica glass merely exposed to water for a time equal to that required for the passage of the crack through the specimen.

The study [5.6] used nuclear reaction analysis (NRA) to determine the concentration profile of water near the fracture surface of the glass. This technique provides a direct chemical analysis of the concentration of H as a function of distance from the fracture surface.

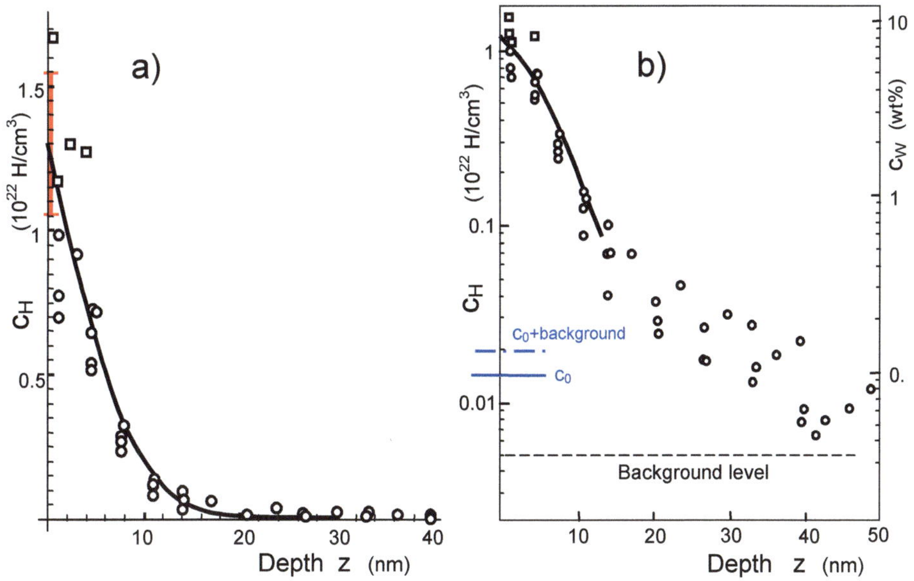

Fig. 5.1 Hydrogen concentration in the surface region of a growing crack via NRA-measurements by Tomozawa et al. [5.6] (circles: from Fig. 4, squares: from Fig. 3 in [5.6]); a) linear ordinate: curve given by eq.(5.2.1), red bar indicates 90%-Confidence Interval of the surface concentration, b) logarithmic ordinate.

In Fig. 5.1 the results from [5.6] are plotted versus depth (circles). We included also the high values from Fig. 3 in [5.6] (indicated in Fig. 5.1 by the squares). It should be noted that the measured data are affected by the finite breadth of the instrument curve. A detailed analysis that includes the effect of the rather large breath of the instrumental is given by Wiederhorn et al. [5.9]. Here we are interested only in the surface values for which the effect of the instrument curve is of minor importance.

In order to allow an approximate extrapolation of the measured data to the surface, we fitted the data c_H in Fig. 5.1a arbitrarily by a complementary error function

$$c_H = c_{H,0}\,\mathrm{erfc}\left[\frac{z}{2b}\right] + \underbrace{0.005\times10^{22}/\mathrm{cm}^3}_{\text{Background}} \tag{5.2.1}$$

taking into account the background level of 0.005 $10^{22}/\mathrm{cm}^3$. The coefficient $c_{H,0}$ is the surface concentration of c_H. By fitting eq.(5.2.1) to the measurements, the best set of parameters was found as

$c_{H,0}$=1.30 [1.055, 1.55] $10^{22}/\mathrm{cm}^3$, b=4.95 [3.28, 6.61] nm (90% CI).

with the 90%-Confidence Intervals in brackets. The red bar in Fig. 5.1a indicates the 90%-CI of the surface concentration. The depth b at which the concentration decreased to ≅50% of the surface value in Fig. 5.1a is about 5 nm.

Using Avogadro's number, N=6.02×10^{23} molecules/mol gives for the water concentration at the surface in molar units:

$$C_w(0) = \frac{c_H}{2N} = 0.0108\ [0.0095, 0.0121]\ (\mathrm{mol/cm}^3) \tag{5.2.2}$$

Defining the mass concentration as the mass of water per mass of glass, the surface value is with ρ=2.2 g/cm³ for the density of silica glass and a molecular mass of water of m_{H2O}=18 g/mol

$$C_w(0) = \frac{m_w}{m_{glass}} = \frac{c_H}{2N}\frac{m_{H_2O}}{\rho} = 8.8\,[7.7, 9.8]\ (\mathrm{wt}\%) \tag{5.2.3}$$

For high water concentrations it is of advantage to use the total mass $m=m_w+m_{glass}$ in the dominator

$$C_w(0) = \frac{m_w}{m_{glass}+m_w} = 8.1\,[7.1, 9.0]\ (\mathrm{wt}\%) \tag{5.2.4}$$

From the result by Zouine et al. [5.10] for unstressed silica surfaces in water, $c_{H,0}$=1.4 10^{20} H-atoms, the water concentration in the absence of stress, $C_{w,0}$, reads

$$C_{w,0} = 0.000116\ (\mathrm{mol/cm}^3) = 0.095\ (\mathrm{wt}\,\%) \tag{5.2.5}$$

The value of c_0 is shown by the blue line in Fig. 5.1b and after including the background by the dash dotted line. From these results, an increase of the water concentration by a factor of $C_w(0)/C_{w,0}\cong 93$ has to be concluded. The right ordinate in Fig. 5.1b represents the total water concentration.

5.2.2 Computation of the equilibrium constant

In molar units, the total water concentration is given by

$$C_w = C + \tfrac{1}{2}S = C(1+\tfrac{1}{2}k) \tag{5.2.6}$$

where the quantity k is the equilibrium constant given by eq.(5.1.6). For the case of saturation, the surface concentration of molecular water, $C(0)$ is generally *assumed* to depend only on the water vapour pressure in the environment, $C(0)=C_0$=const. This leads to

$$C_w = C_0 + \tfrac{1}{2}S = C_0(1+\tfrac{1}{2}k) \tag{5.2.7}$$

In the absence of stresses, the equilibrium constant may be denoted as k_0. In this case, the total water concentration can be written

$$C_{w,0} = C_0 + \tfrac{1}{2}S_0 = C_0(1+\tfrac{1}{2}k_0) \tag{5.2.8}$$

again with subscript "0" for disappearing stress.

The ratio of total water concentrations under stress and in the absence of stresses is

$$\frac{C_w}{C_{w,0}} = \frac{1+\tfrac{1}{2}k}{1+\tfrac{1}{2}k_0} \tag{5.2.9}$$

Since the left-hand side is known from experimental data, the stress-enhanced equilibrium constant can be computed as

$$k = (2+k_0)\frac{C_w}{C_{w,0}} - 2 \tag{5.2.10}$$

Experimental results on equilibrium ratios from literature were expressed in [5.5] for the temperature range of 90°C≤θ≤350°C by the empirical relation (1.1.3).
For room temperature, we obtain by extrapolation to 23°C (T=296°K)

$$k_0(23°C) = 0.41$$

The experimental data for $C_w/C_{w,0}$ in eq.(5.2.10) yield

$$k \cong 222 \Rightarrow \frac{k}{k_0} \cong 540 \tag{5.2.11}$$

an enormous increase of the equilibrium ratio by the stress field near the crack tip. The result clearly illustrates the strong effect of tensile stresses. The equilibrium of the reaction (5.1.1) is strongly shifted to the right side. It is evident that the strong increase of the equilibrium constant must be caused by a positive reaction volume $\Delta\overline{V}>0$ in (5.1.5) as was derived in [5.11]. From eqs.(5.1.5) and (5.1.6) it follows simply

$$\frac{S}{S_0} = \frac{k}{k_0} \tag{5.2.12}$$

The hydroxyl concentrations from eqs.(5.2.3) and (5.2.5) are

$$S(0) = \tfrac{17}{18} C_w(0) / (\tfrac{1}{2} + \tfrac{1}{k}) = 16.5[14.4, 18.4] \ \ (\text{wt\%}) \tag{5.2.13}$$

and

$$S_0 = \tfrac{17}{18} C_{w,0} / (\tfrac{1}{2} + \tfrac{1}{k_0}) = 0.0368 \ \ (\text{wt\%}) \tag{5.2.14}$$

The factor 17/18 is the ratio of the mole masses of hydroxyl (17 g/mol) and molecular water (18 g/mol). In [5.12] we derived from the spherical pore model a maximum possible hydroxyl concentration of S_{max}=15.7 [13.3, 19.2] (wt%). The 90%-confidence intervals of $S(0)$ and S_{max} totally overlap confirming the expectaion that the hydroxyl concentration at the crack tip is the maximal possible one.

5.2.3 Bending tests by Agarwal et al.

In this section, we use the experiments of Agarwal et al. [5.3] to derive information on stress-enhanced swelling of silica glass. Figure 5.2 from Agarwal et al. [5.3] shows the concentration of hydrogen atoms, H, in thin flexure specimens of silica glass, exposed to water vapour at 350 Torr and 650°C. The maximum bending stress in the specimens was ±45 MPa. We evaluated the data in Fig. 5.2 for the times, t=10 h and 20 h. As indicated by the work of Davis and Tomozawa [5.13], water and hence hydrogen at 650°C is present primarily in the form of ≡SiOH groups.

The concentration of H in atoms per cm^3 of glass in Fig. 5.2 was obtained by the technique of Nuclear Reaction Analysis (NRA) [5.3, 5.14]. The OH concentration, *S*, in gm of OH per gm of silica glass, was calculated from the concentration of hydrogen atoms, H, using the following relation:

$$S(gm\,OH\,/\,gm\ silica\ glass) = \frac{H\,/\,cm^3 \times 17 g_{\mathrm{OH}}\,/\,mol}{6.02\times 10^{23}\,/\,mol\cdot 2.2 g_{glass}\,/\,cm^3}. \qquad (5.2.15)$$

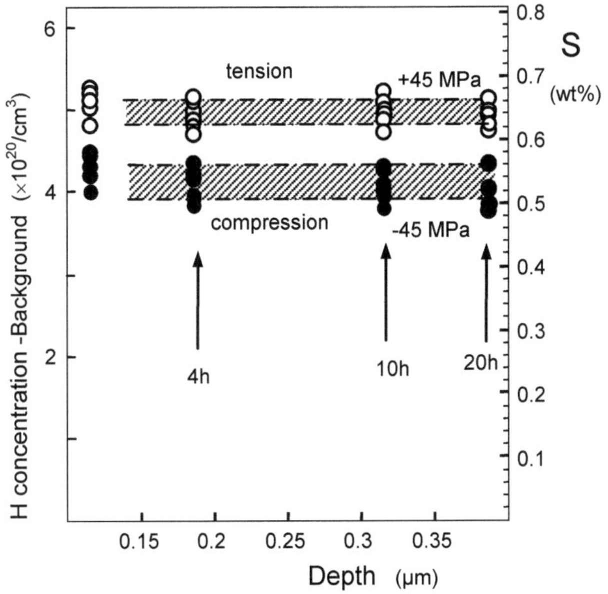

Fig. 5.2 Hydroxyl water concentration below the surface of bent silica glass specimens: water vapour pressure 355 Torr; temperature 650°C; exposure time 4 h, 10 h and 20 h, taken from Agarwal et al. [5.3], shaded areas: ±1 SD.

The 10 data points in the **tensile** region at +45 MPa showed a mean value of $S^{+45} = 0.638$ wt% with a standard deviation of SD = 0.022 wt%. The standard error (SE) of the mean is SE = 0.0071 wt%, or $S^{+45} =$ (0.638±0.0071) wt%.

The 11 data points in the **compressive** side showed a mean value of S^{-45} = 0.539 wt%, SD = 0.0173 wt%, and SE = 0.00523 wt%, hence, S^{-45} = (0.539±0.00523) wt%.

Using the error propagation law gives for the **ratio** of the two concentrations

$$\frac{S^{+45}}{S^{-45}}=\frac{0.638\pm0.0071}{0.539\pm0.00523}=\frac{0.638\pm1.1\%}{0.539\pm0.97\%}=1.184\pm1.5\% \tag{5.2.16}$$

The hydroxyl concentration at the neutral position of the cross-section is determined as the geometric mean of S^{+45} and S^{-45}, i.e., $(S^{+45}\times S^{-45})^{1/2}$. Since Agarwal et al. [5.3] mention that *the hydrogen concentration of the unstressed samples fell between the concentrations of samples under tension and compression*, we will identify this mean concentration with the concentration in the absence of an applied stress:

$$S(0)\cong 0.586\,wt\%\,. \tag{5.2.17}$$

The concentration of H was measured by Nuclear Reaction Analysis. Normally, H is divided between H_2O and ≡SiOH, but at 650°C almost all of the H is present in the form of ≡SiOH. Hence, the measured H concentration can be considered equal to the concentration of ≡SiOH.

In [5.3] the concentration of ≡SiOH was measured additionally by IR. These measurements showed S^+/S^- = 1.19 that is in good agreement with S^+/S^- = 1.184 from the NRA measurements. Within experimental scatter, the results were the same. The absolute value of S_0 however depended substantially on which technique was used for the measurement. The NRA measurements were a factor of three times larger than the equivalent IR measurements. The reason for this discrepancy is not yet clear.

The experiments in [5.3] were carried out under uniaxial stresses (45 MPa on the tension and -45 MPa on the compression side). For an evaluation of these results, a relation may be used similar to eq.(5.1.5) that describes $S(\sigma_{appl})$, where $\sigma_{appl}=\pm45$ MPa:

$$S=S_0\exp\left(\frac{\sigma_{appl}\Delta V_{eff}}{RT}\right) \tag{5.2.18}$$

ΔV_{eff} is an effective activation volume that will of course deviate from $\overline{V}_S$ in the eq.(5.1.5). The ratio of the concentrations in tension and compression is consequently

$$\frac{S^+}{S^-}=\exp\left(\frac{2\sigma_{appl}\Delta V_{eff}}{RT}\right) \tag{5.2.19}$$

From the bending experiment with S^{+}/S^{-}=1.184, eq.(5.2.16), we obtain for the case of stress-enhanced swelling and T>450°C

$$\Delta V_{eff} = 14.4\,\mathrm{cm^3/mol} \pm 10\% \tag{5.2.20}$$

References

5.1 H. Le Chatelier, *C.R. Acad. Sci. Paris* **99**(1884), 786.

5.2 J. F. Shackelford, "Gas Solubility in Glasses: Principles and Applications," Int. J. Appl. Glass Sci, **2** [2] (2011) 85–95.

5.3 A. Agarwal, M. Tomozawa, W. A. Lanford, Effect of stress on water diffusion in silica glass at various temperatures, J. Non-Cryst. **167**(1994), 139-148.

5.4 K.B. McAfee, Jr., J. Chem. Phys. **28**(1958) 218.

5.5 S. M. Wiederhorn, F. Yi, D. LaVan, T. Fett, M.J. Hoffmann, Volume Expansion caused by Water Penetration into Silica Glass, J. Am. Ceram. Soc. **98**(2015), 78-89.

5.6. M. Tomozawa, W.-T. Han and W.A. Lanford, "Water Entry into Silica Glass during Slow Crack Growth," J. Am. Ceram. Soc. **74** [10] (1991) 2573-2576.

5.7. C. Janssen, in *Proceedings of Tenth International Congress on Glass, Kyoto, Japan* 1974. (Ceramic Society of Japan, Tokyo, Japan 1974) p. 10.23

5.8. Michalske, T.A., Smith W.L. and Chen, E.P. "Stress Intensity Calibration for the Double Cleavage Drilled Compression Specimen," Eng. Frac. Mech. **45**(1993), 637-42.

5.9 S. M. Wiederhorn, T. Fett, G. Rizzi, M. J. Hoffmann, J.-P. Guin, The Effect of Water Penetration on Crack Growth in Silica Glass, Engng. Fract. Mech. **100**(2013), 3-16

5.10 A. Zouine, O. Dersch, G. Walter and F. Rauch, "Diffusivity and solubility of water in silica glass in the temperature range 23-200°C," *Phys. Chem. Glass: Eur. J. Glass Sci and Tech. Pt. B*, **48** [2] (2007), 85-91.

5.11 K.G. Schell, E.C. Bucharsky, T. Fett, The sign of the reaction volume for the water/silica reaction, Glass Physics and Chemistry, **46** (2020), 49-52.

5.12 T. Fett, G. Schell, Damage in silica by hydroxyl damage, Scientific Working Papers **85**, 2015, ISSN: 2194-1629, Karlsruhe, KIT.

5.13 K.M. Davis, and M. Tomozawa, "An infrared spectroscopic study of water-related species in silica glasses," *J. Non-Cryst. Solids,* **201** 177-198 (1996).

5.14 Helmich, M. Rauch, F., On the mechanism of diffusion of water in silica glass. *Glastech. Ber.*, 1993, **66**, 195–200.

6 Swelling anisotropy

6.1 Motivation for anisotropic swelling

Strongly different stress states can result in the same hydrostatic stress term σ_h. An example for this fact is illustrated in Fig. 6.1 where the hydrostatic stress disappears completely, $\sigma_h=0$. This is trivial when all stress components disappear, Fig. 6.1a. In the case of Fig. 6.1b where the transversal stresses are half of the longitudinal stress with inverse sign, it is again $\sigma_h=0$ according to $\sigma_h=(\sigma_x+\sigma_y+\sigma_z)/3$.

The same holds for torsion loading, $\sigma_y=-\sigma_z$, Fig. 6.1b'. It is hardly imaginable that high tensile stresses *in* bond direction should have no effect on bond fracture. It has to be expected that the reaction in tensile direction should produce swelling even for the hydrostatic stress state $\sigma_h=0$.

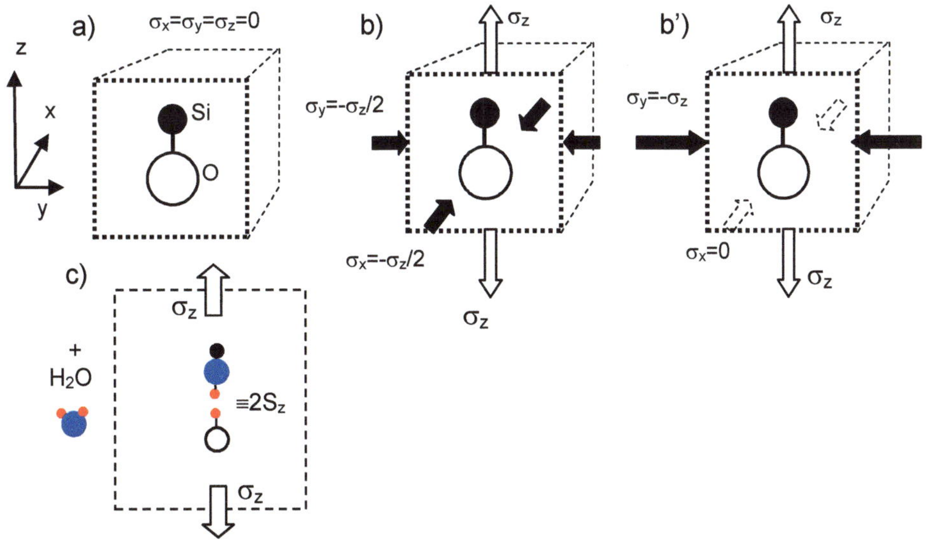

Fig. 6.1 Three stress states acting on a Si-O-bond exhibiting the same disappearing hydrostatic stress $\sigma_h=0$: a) isotropic swelling is expected, for b) and b') swelling in *z*-direction should dominate, (case b' describes torsion loading), c) water molecule reacting with a stressed bond oriented in *z*-direction, producing two hydroxyls, $2S_z$.

Figure 6.2 shows the load-displacement curves for bond breaking by reaction (1.1.1). Perpendicular to the loading direction, Fig. 6.2a, the necessary energy for splitting the bond is U that must be delivered by thermal fluctuations.

The portion N_x (subscript x means the bonds oriented in x-direction in which no stress acts) of opened bonds results from the Boltzmann equation as

$$N_x \propto \exp\left(-\frac{U}{RT}\right) \tag{6.1.1}$$

The same equation holds for the y-direction

$$N_y \propto \exp\left(-\frac{U}{RT}\right) \tag{6.1.2}$$

The potential barrier U belongs to the case when no stresses are present or when the bond is perpendicular to the stress, Fig. 6.2a.

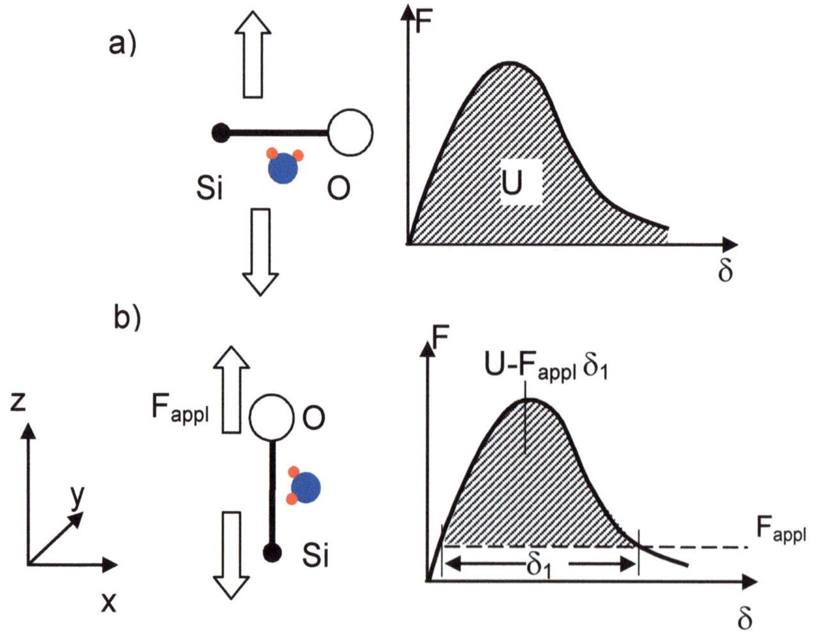

Fig. 6.2 Energy conditions for bond breaking.

Under action of a force *F* in bond-direction, Fig. 6.2b, the potential barrier is reduced to $U\text{-}F\delta_1$. Consequently, the occurrence of opened bonds in *z*-direction is increased

$$N_z \propto \exp\left(-\frac{U-\sigma_z V}{RT}\right) \tag{6.1.3}$$

where *V* is an activation volume. In the following considerations, it will be shown that the strongest anisotropy effect has to be expected for the torsion loading case.

For the computations on anisotropic swelling let us follow the derivations made in [6.1]. The stress effect on the equilibrium constant for reaction (1.1.1) can be derived from Eq.(6) of the article by de Boer [6.2]. Following the procedure in the review article by Hamann [6.3], we obtain the following equation for the equilibrium constant:

$$RT\ln k = \mu_{0,\mathrm{H2O}} - 2\mu_{0,\mathrm{SiOH}} + 2\Delta W + 2\Delta U\,, \tag{6.1.4}$$

where $\mu_{0,H2O}$ and $\mu_{0,SiOH}$ are the chemical potentials for molecular water and hydroxyls, respectively in the unstressed state; ΔW represents the work per mol of SiOH done by the mechanical stresses; and ΔU is the change of the elastic strain energy per mol due to a reduction of material stiffness as a consequence of the water reaction.
The work term ΔW in Eq.(6.1.4) is for loading in 3 directions $\sigma_x, \sigma_y, \sigma_z$

$$\Delta W / V_0 = \sigma_z \varepsilon^{(1)}_{sw,z} + \sigma_y \varepsilon^{(1)}_{sw,y} + \sigma_x \varepsilon^{(1)}_{sw,x} \tag{6.1.5}$$

with the Cartesian components of swelling strains $\varepsilon_{sw,i}$, the glass volume, V_0, conveniently chosen as the molar volume of dry glass, and the superscripts $^{(1)}$ standing for 1 mole of *S*. Since the total swelling volume is in general the sum of the components, it holds for the individual strain components $\varepsilon_{sw,i}^{(1)}$ and the volume swelling strain $\varepsilon_v^{(1)}$ for 1 mole of *S* in the volume V_0

$$\varepsilon^{(1)}_v = \varepsilon^{(1)}_{sw,x} + \varepsilon^{(1)}_{sw,y} + \varepsilon^{(1)}_{sw,z} \tag{6.1.6}$$

The change of strain energy is for uniaxial loading by $\sigma_z = \sigma_{appl}$

$$\Delta U / V_0 = -\tfrac{1}{2}\varepsilon^2_{appl}\Delta E \tag{6.1.7}$$

where ΔE is the change in the Young's modulus as a consequence of the material damage by hydroxyl generation.

The reduction in stiffness results in a change of the quantity of elastically stored energy, which in turn, contributes to the driving force of the reaction [6.4]. Since $\Delta E \leq 0$, the contribution ΔU has the same sign as ΔW. In the evaluation in [6.1], the energy term, ΔU, was neglected assuming that $\Delta E/\Delta S$ is small due to the low hydroxyl concentrations in our calculations. Hence, we use the undamaged Young's modulus in all equations having in mind that the results of our computations are <u>lower limits</u>.

Equations (6.1.4) and (6.1.5) now yield

$$k = k_0 \exp\left[2V_0 \frac{(\sigma_z \varepsilon_{sw,z}^{(1)} + \sigma_y \varepsilon_{sw,y}^{(1)} + \sigma_x \varepsilon_{sw,x}^{(1)})}{RT}\right], \qquad (6.1.8)$$

and, consequently

$$S = S_0 \exp\left[V_0 \frac{(\sigma_z \varepsilon_{sw,z}^{(1)} + \sigma_y \varepsilon_{sw,y}^{(1)} + \sigma_x \varepsilon_{sw,x}^{(1)})}{RT}\right], \qquad (6.1.9)$$

where k_0 includes all of the terms that are independent of stress ($\mu_{0,SiOH}$, $\mu_{0,H2O}$ and $\ln[C]$).

The molar volume for glass is V_0=27.27 cm^3/mol. For one mole of hydroxyl with the mass m_{OH}=17g/mole in the volume V_0 we obtain the hydroxyl concentration in weight units (mass S/mass SiO_2)

$$S^{(1)} = \frac{m_{\mathrm{OH}}}{M_{\mathrm{glass}}} = \frac{17\ \mathrm{g/mole}}{60\ \mathrm{g/mole}} = 0.283 \qquad (6.1.10)$$

with the related volume strain according to eq.(2.1.11)

$$\varepsilon_v^{(1)} = \kappa S^{(1)} = 0.274 \qquad (6.1.11)$$

Then it holds for $V_0 \varepsilon_v^{(1)}$

$$V_0 \varepsilon_v^{(1)} = 27.27 \times 0.274 \cong 7.5\ \mathrm{cm^3/mole} \qquad (6.1.12)$$

6.2 Pure tensile loading

Under purely uniaxial tensile stresses $\sigma_z = \sigma_{appl} > 0$, $\sigma_x = \sigma_y = 0$, it results

$$S = S_0 \exp\left[V_0 \frac{\sigma_{appl}\, \varepsilon_{sw,z}^{(1)}}{RT} \right] \tag{6.2.1}$$

or rewritten

$$S = S_0 \exp\left[V_0 \frac{\sigma_{appl}\, \varepsilon_{\mathrm{v}}^{(1)}}{RT} \frac{\varepsilon_{sw,z}^{(1)}}{\varepsilon_{\mathrm{v}}^{(1)}} \right] = S_0 \exp\left[\frac{\sigma_{appl}}{RT} \frac{\varepsilon_{sw,z}^{(1)}}{\varepsilon_{\mathrm{v}}^{(1)}} \times 7.5\,\mathrm{cm^3/mol} \right] \tag{6.2.2}$$

by introducing eq.(6.1.12) in (6.2.1).

The experimentally found dependency S=f(σ_{appl}) in Fig. 5.2 must agree with the theoretical relation eq.(6.2.2), i.e. it must hold for uniaxial loading $\sigma_{appl}=\sigma_z$:

$$\frac{\sigma_{appl} \Delta V_{eff}}{RT} = \frac{\sigma_{appl}}{RT} \frac{\varepsilon_{sw,z}^{(1)}}{\varepsilon_{\mathrm{v}}^{(1)}} \times 7.5\,\mathrm{cm^3/mol}$$
$$\Rightarrow \quad \frac{\varepsilon_{sw,z}^{(1)}}{\varepsilon_{\mathrm{v}}^{(1)}} = \frac{14.4\,\mathrm{cm^3/mol}}{7.5\,\mathrm{cm^3/mol}} = 1.92 \tag{6.2.3}$$

Consequently, we obtain:

$$\varepsilon_{sw,z}/\varepsilon_{\mathrm{v}} = 1.92. \tag{6.2.4}$$

and for the other components [6.1]:

$$\varepsilon_v = \varepsilon_{sw,z} + 2\varepsilon_{sw,x} \Rightarrow \varepsilon_{sw,x} = \tfrac{1}{2}(\varepsilon_v - \varepsilon_{sw,z}) \Rightarrow \frac{\varepsilon_{sw,x}}{\varepsilon_v} = \frac{\varepsilon_{sw,y}}{\varepsilon_v} = -0.46 \tag{6.2.5}$$

Finally, it holds for the uniaxial load

$$S = S_0 \exp\left[\frac{\sigma_{appl}\, \Delta V_{eff}}{RT} \right] \tag{6.2.6}$$

6.3 Swelling stresses

Since in water diffusion layers the swelling strains $\varepsilon_{sw,i}$ are mostly present in thin surface layers, free expansion is not possible due to the bulk material that remains free of water. Due to this restriction, the swelling strains must cause swelling stresses $\sigma_{sw,z}$ and $\sigma_{sw,y}$ in the surface layer. These swelling stresses are responsible for the fact that even under purely uniaxial loading a biaxial stress state prevails near the surface.

Then, the total stress in the surface layer is

$$\sigma_z = \sigma_{appl,z} + \sigma_{sw,z}\,,\quad \sigma_y = \sigma_{appl,y} + \sigma_{sw,y} \tag{6.3.1}$$

For biaxial stresses we get from (6.1.9)

$$S = S_0 \exp\left[V_0 \frac{(\sigma_z \varepsilon^{(1)}_{sw,z} + \sigma_y \varepsilon^{(1)}_{sw,y})}{RT} \right], \tag{6.3.2}$$

In our special application an applied stress in *y*-direction is not present, $\sigma_{appl,y} = 0$, so that eq.(6.3.2) reads after renaming $\sigma_{appl,z} = \sigma_{appl}$ and use of eq.(6.2.5)

$$S = S_0 \exp\left[\left(\sigma_{appl} + \sigma_{sw,z} + \frac{-0.46}{1.92} \sigma_{sw,y} \right) \frac{\Delta V_{eff}}{RT} \right] \tag{6.3.3}$$

Hooke's law gives the total strains in a thin surface layer as the sum of the swelling and the elastic strains

$$\varepsilon_{total,z} = \varepsilon_{sw,z} + \frac{1}{E}(\sigma_{sw,z} - \nu \sigma_{sw,y}) \tag{6.3.4a}$$

$$\varepsilon_{total,y} = \varepsilon_{sw,y} + \frac{1}{E}(\sigma_{sw,y} - \nu \sigma_{sw,z}) \tag{6.3.4b}$$

With $\varepsilon_{total,z} = \varepsilon_{total,y} = 0$, equations (6.3.4a) and (6.3.4b) are simultaneous equations with two unknowns: $\sigma_{sw,z}$ and $\sigma_{sw,y}$. The solution of this system of equations is

$$\sigma_{sw,z} = -\frac{E}{1-\nu^2}(\varepsilon_{sw,z} + \nu \varepsilon_{sw,y}) \tag{6.3.5a}$$

$$\sigma_{sw,y} = -\frac{E}{1-\nu^2}(\varepsilon_{sw,y} + \nu\varepsilon_{sw,z}) \tag{6.3.5b}$$

Introducing eqs.(6.3.5a, 6.3.5b) and (6.2.5) results in the exponent of eq.(6.3.3)

$$\sigma_{sw,z} + \frac{\varepsilon_{sw,y}}{\varepsilon_{sw,z}}\sigma_{sw,y} = -\frac{\kappa S E}{4(1-\nu^2)}\left(\frac{\varepsilon_v}{\varepsilon_{sw,z}} + \frac{\varepsilon_{sw,z}}{\varepsilon_v}(5-4\nu) - 2(1-2\nu)\right) \tag{6.3.6}$$

so that eq.(6.3.2) can be written as an implicit equation with respect to S

$$S = S_0 \exp\left[(\sigma_{appl} + \beta S)\frac{\Delta V_{eff}}{RT}\right], \tag{6.3.7}$$

The abbreviation β reads by using the material data E=72GPa, ν=0.17, and κ=0.97

$$\beta = -\frac{\kappa E}{4(1-\nu^2)}\left(\frac{\varepsilon_v}{\varepsilon_{sw,z}} + \frac{\varepsilon_{sw,z}}{\varepsilon_v}(5-4\nu) - 2(1-2\nu)\right) = -135 \text{ GPa} \tag{6.3.8}$$

The explicit form of eq.(6.3.7) is:

$$S = -\frac{RT}{\Delta V_{eff}\,\beta}\text{PLog}\left[-\frac{\Delta V_{eff}}{RT}\beta S_0 \exp\left(\sigma_{appl}\frac{\Delta V_{eff}}{RT}\right)\right], \tag{6.3.7a}$$

where PLog stands for the *product-logarithm function* (see e.g. [6.5]):

The axial swelling stress results from eqs.(6.3.5a) and (6.1.6) as

$$\sigma_{sw,z} = -\kappa\frac{E}{1-\nu^2}\left(\alpha\left(1-\frac{\nu}{2}\right)+\frac{\nu}{2}\right)\times S \cong -\eta_z S, \quad \eta_z = 133 \text{ GPa} \tag{6.3.9}$$

with $\alpha=\varepsilon_{sw,z}/\varepsilon_v$ as an anisotropy coefficient.

References

6.1 S. M. Wiederhorn, G. Rizzi, S. Wagner, M. J. Hoffmann, T. Fett, Stress-Enhanced Swelling of Silica: Effect on Strength, J. Am. Ceram. Soc. **99**(2016), 2956-63.

6.2 R.B de Boer, On the thermodynamics of pressure solution – interaction between chemical and mechanical forces, Geochimica et Cosmochimica Acta, **41** (1977) 249-256

6.3 S.D. Hamann, Chemical equilibria in condensed systems, In: High Pressure Physics and Chemistry, Ed. R.S. Bradley, p. 131, 1963, Academic Press, London.

6.4 T. Zhu, J. Li, X. Lin, S. Yip, Stress-dependent molecular pathways of silica-water reaction, J. of Mechanics and Physics of solids, **53**(2005), 1597-16211.

6.5 R. Corless, G. Gonnet, D. Hare, D. Jeffrey and D. Knuth, "On the Lambert W Function," Advances in Computational Mathematics, **5**(1996), 329-359.

7 Strength of silica fibers

7.1 Tensile strength of fibers

Soaking of silica under tensile loading improves strength. To the knowledge of the autors, this increase was qualitatively found first by Proctor et al. (Section 3.3 in [7.1]). In a recent study by Lezzi et al. [7.2], uniaxial tensile stresses, 1 GPa to 3 GPa, were applied to silica fibers, which were then exposed to water vapour, 6 Torr vapour pressure. After 1 min of exposure, followed by cooling to room temperature, fiber strengths were measured in normal lab air at 20°C. Figure 7.1a shows the strengths as a function of soaking temperature for different tensile stresses σ_{appl}. Perpendicular bars show ± 1 standard deviation.

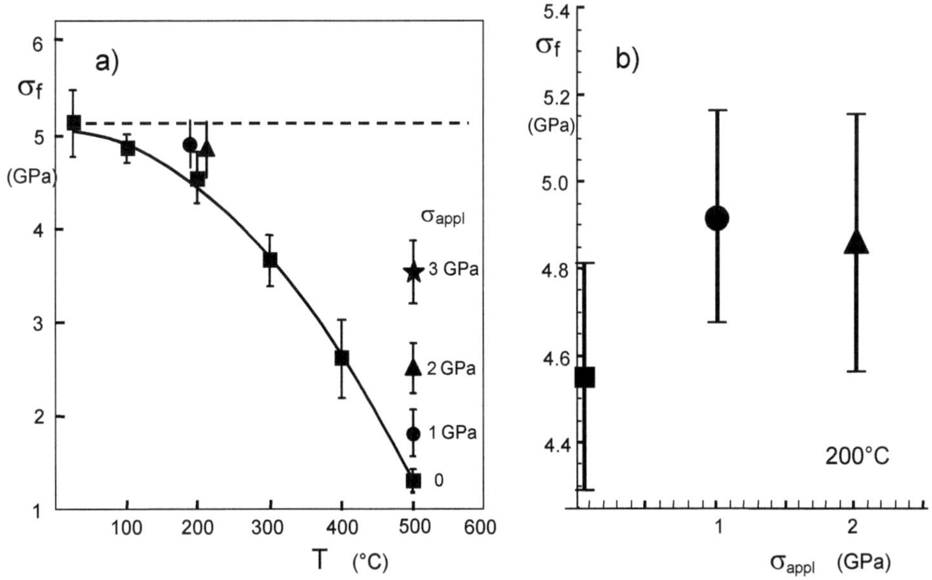

Fig. 7.1 a) Strength measurements on silica fibers thermally treated for 60 s in nitrogen atmosphere with a partial water pressure of 6 Torr under different tensile stresses; b) strength at 200°C vs. applied stress during thermal treatment (results from Lezzi et al. [7.2]).

In the evaluations of strength we concentrate on 200°C and 500°C since at these temperatures the best data base is reported.

In Fig. 7.1b the 200°C-data are plotted in some more detail. For 500°C, Fig. 7.2a shows the increase of critical stresses, i.e. of strength, $\Delta\sigma_c$,

$$\Delta\sigma_c = \sigma_c(\sigma_{appl}) - \sigma_c(0) \tag{7.1.1}$$

as a function of the applied stress where $\sigma_c(0)$ is the strength of unstressed fibers. The circles indicate the median values and the bars represent ± one standard deviation.

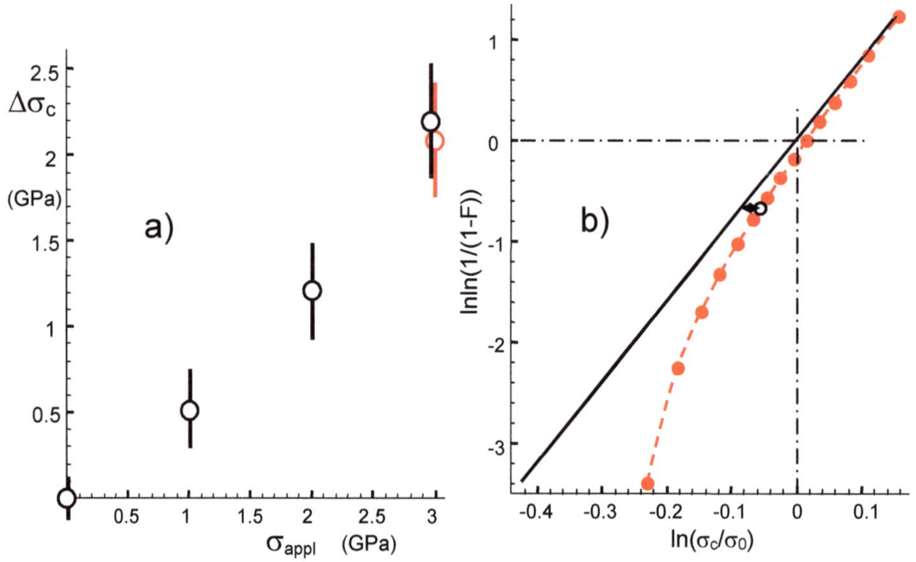

Fig. 7.2 a) Strength results on silica fibers at 500°C, Red data: corrected strength; b) Effect of 2 test specimens cracked during soaking at 3GPa. These specimens were ignored in strength evaluation by Lezzi et al. [7.2].

In our opinion the measured strength data at 3 GPa are slightly too high [7.3]. The authors mention that at the highest load some fibers (about 10%) fractured already during soaking. Under constant test conditions, these fibers, which were inevitably excluded from the later strength tests, must have had the largest defects. This corresponds to the well-known problem of a proof-test preceding the strength measurements. In principle, one has to evaluate a strength distribution truncated at the lower strength values [7.4]. The way of evaluation has of course an effect on the strength of the surviving specimens and may be illustrated in Fig. 7.2b for a Weibull modulus of m=8. The black curve represents the Weibull distribution for the case that no specimen would have failed during the soaking procedure. For the case that two specimens might have failed during soaking, the failure probability has to be computed on the basis of 17 specimens and not on the basis of 15 specimens. The true characteristic strength σ_0 is 1.5% smaller than the value reported in [7.2] and the median value by 3% smaller. The change in $\Delta\sigma_c$ is then slightly reduced by 4.2%×$\Delta\sigma_c \approx -95$ MPa. Figure 7.2a shows the results after re-evaluation as the red data point together with the original data. The slight shift in σ_{appl} is for an improved distinction.

7.2 Strength predictions including swelling

7.2.1 Effect of global swelling stresses

The swelling stresses in Section 3 were obtained under condition of the absence of external loading. So we could always assume *isotropic swelling*, *i.e.* strains were assumed to be identical in all directions. As this might not be the case for some recent high tensile stress experiments carried out on silica glass [7.2], a reconsideration of the equilibrium conditions is necessary according to Section 6.3 for a valid determination of the effect of applied stress on the equilibrium constant for reaction (1.1.1). In this section information on stress-enhanced swelling of silica glass is used to discuss the experiments of Lezzi et al. [7.2] in terms of water penetration and the consequent swelling of silica glass.

In the high-stress soaking tests on silica fibers by Lezzi et al. [7.2], the externally applied loads were uniaxial. Therefore, we concentrate here also on the uniaxial description of the effect of stress on ≡SiOH concentration in the glass by applying σ_z exclusively. As outlined in Section 6.3, the swelling stress in axial direction is according to eq.(6.3.9) and [7.3]

$$\sigma_{sw,z} \cong -133 \times S \quad \text{(GPa)} \tag{7.2.1}$$

The strength increase with respect to the strength of unstressed fibers is

$$\Delta\sigma_c = -\sigma_{sw,z}(\sigma_{appl}) + \sigma_{sw,z}(0) \qquad (7.2.2)$$

where $\sigma_{sw,z}(0)$ is the swelling stress in the absence of an externally applied stress, $\sigma_{appl}=0$.

Figure 7.3 shows the hydroxyl concentration at a water vapour pressure of 355 Torr as a function of temperature, reported by Davis and Tomozawa [7.5]. The circles in Fig. 7.3 indicate data from IR-analyses. The red triangle is from the IR-data in reference [7.6], whereas the blue triangle is from the NRA analysis in reference [7.8]. The blue square is an extrapolated result, based on the slope of the Davis and Tomozawa data [7.5] and the position of the blue triangle.

Assuming that the ratio $S_{NR}/S_{IR}\approx 3$ is constant from 650°C to 500°C, we can use the general temperature trend shown in Fig. 7.3 to scale the value of S from 650°C to 500°C. From the result $S(0)$ = 0.586 wt% at 650°C, we obtain from Fig. 7.3 for 500°C: $S(0)=0.586\times1.75=1.026$ wt%. This concentration results in the concentration of $S_{0,500°C,355Torr}$= 1.28 wt%. Finally, the concentrations, S_0, were transformed from 355 Torr to 6 Torr water vapour pressure (as was used by Lezzi et al. [7.2]) via

$$S_{6\,\mathrm{Torr}} = S_{355\,\mathrm{Torr}}\sqrt{\frac{6}{355}} \qquad (7.2.3)$$

$$\Rightarrow S_{0,500°C,\,6Torr} \cong 0.167\ \mathrm{wt.\%}$$

Table 7.1 compiles the S-concentrations derived.

Temperature	$S(0)$ at 355Torr (wt%)		$S(0)$ from Fig. 7.3 (355 Torr) (wt%)		S_0 at 6 Torr	
	NRA	IR	NRA	IR	NRA	IR
650°C	0.586	0.19			-	-
500°C			1.026	0.333	0.167	0.047

Table 7.1 Hydroxyl concentrations in wt-% used for the strength predictions; $S(0)$ at 650°C represents the hydroxyl content in the neutral axis of the bending tests by Agarwal et al. [7.6]. These results were used in Fig. 7.3.

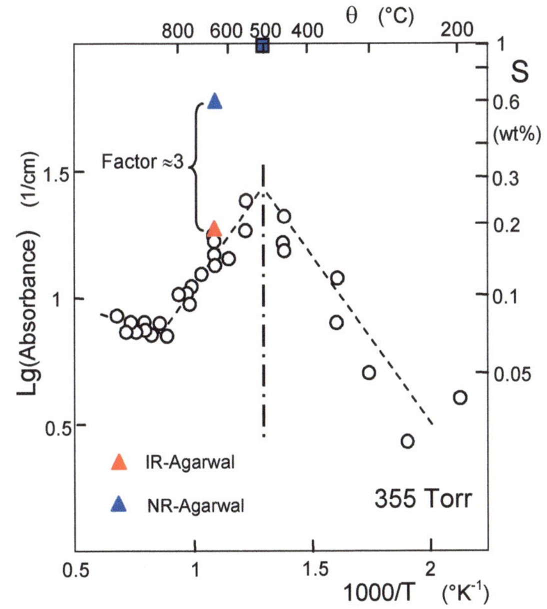

Fig. 7.3 The hydroxyl concentration S in silica, reported by Davis and Tomozawa [7.5] at a water vapour pressure of $p = 355$ Torr (circles); the red triangle indicates the result of the IR-evaluation in [7.6]. The blue triangle represents the concentration obtained by an NRA analysis in [7.6].

7.2.2 Application on strength

Strength at 500°C: In a recent study by Lezzi et al. [7.2], uniaxial tensile stresses, 1 GPa to 3 GPa, were applied to silica fibers; the fibers were then exposed to water vapour, 6 Torr vapour pressure at 500°C. After 1 min of exposure, followed by cooling to room temperature, fiber strengths were measured in normal lab air at 20°C. No measurements of hydroxyl water by the NRA-method are available for that experiment for comparison with our computations. However, Agarwal et al. [7.6] measured the hydroxyl distribution at 650°C by both NRA and IR−analysis. Therefore, a comparison of the two measurements is possible at that temperature. From the IR measurements in Fig. 7b of their paper, we get the following surface values from the bent specimen:

$$S^{45} = 20.3 \text{ Abs/cm},\ S^{-45} = 17 \text{ Abs/cm},\ S(0) = 18.6 \text{ Abs/cm}$$

(1 Abs unit/cm ≅ 0.1 wt% OH). The ratio of the concentrations at the two surfaces, $S^+/S^- \cong 1.19$, is essentially the same as found for the NRA-results presented above, $S^+/S^- \cong$

1.184. By contrast, the value of $S(0)$ for the unstressed specimen differs from the NRA method by a factor of about three. For the IR measurement $S(0) \cong 0.19$ wt%, whereas for the NRA measurement $S(0) \cong 0.586$ wt%.

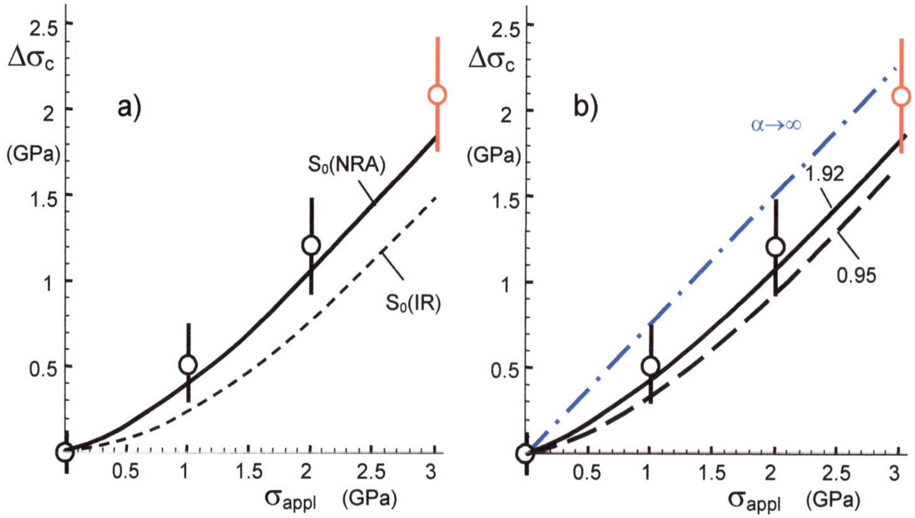

Fig. 7.4 a) Strength increase at 500°C vs. applied stress during thermal treatment for anisotropic swelling; Circles: results from Lezzi et al., bars: ±1 SD, curves: Effect of global swelling stresses $-\sigma_{sw}$ computed with the product of $\alpha\, V_0\varepsilon^{(1)}_v$=14.4 cm³/mole, α=1.92 and S_0 from NR-analysis (solid curve) and from IR-data (dashed curve), b) sensitivity of strength prediction on the anisotropy coefficient α. For the red data point, see Section 7.1.

Figure 7.4 shows the strength increase measured by Lezzi et al. [7.2] as a function of the applied stress. The circles indicate the median values and the perpendicular bars ±1 standard deviation. As already mentioned, the measured strength data at 3 GPa given in [7.2] are slightly too high. A correction of this data point is given in Section 7.1, resulting in the red circle.

The swelling stresses are plotted in Fig. 7.4a by the solid curve for the S-measurement via NRA and the dashed curve for the IR measurement; both values were calculated from Agarwal et al. [7.7] and are indicated in Fig. 7.3 by the triangles. In Fig 7.4b the strength predictions via eqs.(7.1.1) and (7.1.2) are shown for different values of the anisotropy coefficient α. It can be stated that the influence of α is not strong.

Strength at 200°C: Figure 7.1b shows the strengths as a function of the applied stress during soaking at 200°C. For the prediction of strength a modification of the derivation in [7.3] is necessary. In the low-temperature region of θ<500°C, the equilibrium constant of reaction (1.1.1) is

$$k_1 = \frac{S}{C} \tag{7.2.4}$$

The procedure in the article by Hamann [7.8] results in the equilibrium constant:

$$RT \ln k_1 = \mu_{0,\mathrm{H2O}} - \mu_{0,\mathrm{SiOH}} + \Delta W + \Delta U \, , \tag{7.2.5}$$

The equilibrium constant under stress is

$$k_1 = k_{1,0} \exp\left[V_0 \frac{(\sigma_z \varepsilon_{sw,z}^{(1)} + \sigma_y \varepsilon_{sw,y}^{(1)} + \sigma_x \varepsilon_{sw,x}^{(1)})}{RT} \right], \tag{7.2.6}$$

Combining eqs.(7.2.6) and (7.2.4) yields the stress-enhanced hydroxyl concentration

$$S = S_0 \exp\left[V_0 \frac{(\sigma_z \varepsilon_{sw,z}^{(1)} + \sigma_y \varepsilon_{sw,y}^{(1)} + \sigma_x \varepsilon_{sw,x}^{(1)})}{RT} \right] \tag{7.2.7}$$

This is the same relation as was derived in [7.3] for the high-temperature region of θ≥500°C. This makes it possible to take over the main results from [7.3].The hydroxyl concentration data in mass units under water saturation pressure are approximated in Section 4 as

$$S_{\mathrm{sat}} \cong 0.000265 \exp(0.0143\,\theta) \tag{7.2.8}$$

with the temperature θ in °C. For 200°C we obtain $S_{\mathrm{sat}} \cong 0.5$ wt%. This concentration is of course not the surface value in a stress-free state since this concentration causes swelling stresses and is therefore the concentration under a compressive swelling stress. In the absence of externally applied uniaxial loading, the swelling is isotropic. This implies (Section 6.3)

$$\alpha = 1/3 \text{ and } \beta_{1/3} = -56.1 \text{ GPa}$$

The hydroxyl concentration S_0 results from

$$S_0 = S_{sat} \exp\left[-\beta_{1/3} S_{sat} \frac{\frac{1}{3} V_0 \varepsilon_v^{(1)}}{RT} \right] \cong 0.6 \text{ wt\%} \tag{7.2.9}$$

Since saturation pressure at 200°C is 1554 kPa, the concentration for p=6 Torr is

$$S_{0,\,6Torr} \cong 3.1 \text{ ppm}$$

Introducing this value in eq.(7.2.7) gives the curve $S(\sigma_{appl})$ shown in Fig. 7.5a by the right ordinate.

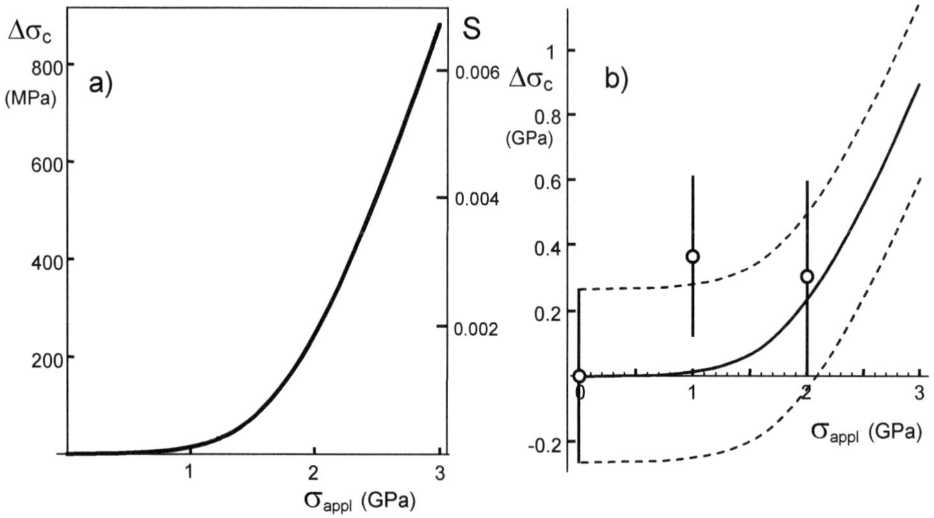

Fig. 7.5 a) Effect of soaking under stress for 6 Torr/200°C: left ordinate: strength increase, right ordinate: related hydroxyl concentration, b) prediction of strength increase (curve).

The axial swelling stress was obtained from eq.(7.1.1) and the strength increase from eq.(7.2.2). For a stress of 1 GPa there is not any strength increase visible and even for 2 GPa the increase of tensile strength is only ≈250 MPa. For a stress of 3 GPa we can expect a strength increase of about 0.9 GPa and a hydroxyl concentration of $S \approx 0.7$ wt-%. Figure 7.5b finally compares the measurements by Lezzi et al. [7.2] from Fig. 7.1a together with the predictions. The measurements are again reported by the solid symbols, the predictions by the open ones. Figure 7.6 shows the strength predictions for the different applied tensile stresses together with the measured data as a function of soaking temperature.

In the preceding computations only the effect of global swelling in the diffusion layer was taken into account. For the alternate strength description by crack-like defects, it is of importance to compare the size of the surface cracks, *a*, with the thickness of the water-affected zone, *b*, given by eq.(2.2.2). From Zouine et al. [7.7] we obtain D_e=8×10^{-13} cm^2/s. Consequently: $b\approx$ 70 nm.

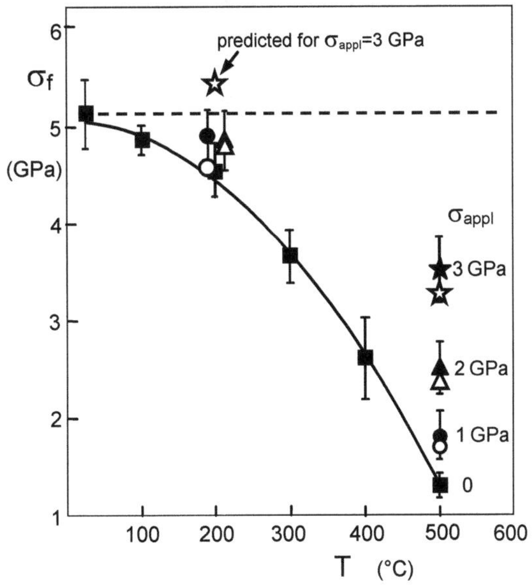

Fig. 7.6 Predictions (open symbols) compared with the strength measurements of Fig. 7.1a.

The crack depths can be concluded from strength results. Assuming the surface cracks as edge cracks, the crack depth is given by the fracture condition $K=K_{Ic}$ and

$$K_{Ic} = \sigma_f \, F \sqrt{\pi \, a} \tag{7.2.10}$$

with the geometric function F=1.122. For K_{Ic}=0.8 MPa√m [7.9] and σ_f=4.6 GPa one obtains from (7.2.10):

$$a=7.6 \text{ nm}$$

Since $a \ll b$, cracks are fully embedded in the water-affected zone.

7.3 Effect of local swelling at surface cracks

During the soaking treatment the crack-tip region is surrounded by a swelling zone. The shielding stress intensity factor caused by such a zone is given by [7.10]

$$K_{sh} = -\psi \frac{\varepsilon_v E}{1-\nu}\sqrt{\omega} \tag{7.3.1}$$

where ω is the height of the swelling zone. This shielding stress intensity factor causes an increase of inert strength $\Delta\sigma_c$

$$\Delta\sigma_c = \frac{-K_{sh}}{\sqrt{\pi a_0}} = \psi \frac{\varepsilon_v E}{(1-\nu)\sqrt{\pi}}\sqrt{\frac{\omega}{a_0}}\,. \tag{7.3.2}$$

The zone of an arrested crack is heart-shaped resulting in the coefficient ψ=0. After high-temperature soaking, these swelling zones are maintained at room temperature where the strength tests are performed.

In the strength tests, the crack must extend and escape of the initial swelling zone. Consequently, the coeffient ψ in eq.(7.3.1) must become dependent on the crack extension Δa in the strength test. By application of the McMeeking and Evans [7.10] procedure as well as by FE-analysis, ψ was computed in [7.11]. The red curve in Fig. 7.7 represents analytical results for a heart-shaped swelling zone we got by application of the McMeeking and Evans [7.10] procedure (see also Appendix A1). A maximum of ψ is reached at $\Delta a/\omega = 8/\sqrt{27} \cong 1.54$ with ψ≈0.3. In the case of a grown crack with extended swelling zone (black curve) it results ψ≈0.33. In order to evaluate eq.(7.3.2), the zone height ω has to be known.

For *room temperature* Lechenault et al. [7.12] used a neutron reflection technique to measure the penetration of heavy water (deuterium oxide) into the silica glass. The work of Lechenault *et al.* [7.12] predicts the water profiles for crack velocities of ≈ 10^{-8} m/s and ≈ $4\text{x}10^{-6}$ m/s. An average zone height of $\omega \cong 6.5$ nm may be concluded for room temperature. This value is a lower limit for high-temperature applications since at 500°C the diffusivity is increased and increased zone heights have to be expected.

The negative intrinsic stress intensity factor causes an increased externally applied stress intensity factor for crack extension. Consequently, the strength must increase.

Due to the water content in the crack-tip region, the Young's module is reduced. Consequently, eq.(7.3.1) must overestimate the shielding effect and (7.3.2) the strength increase.

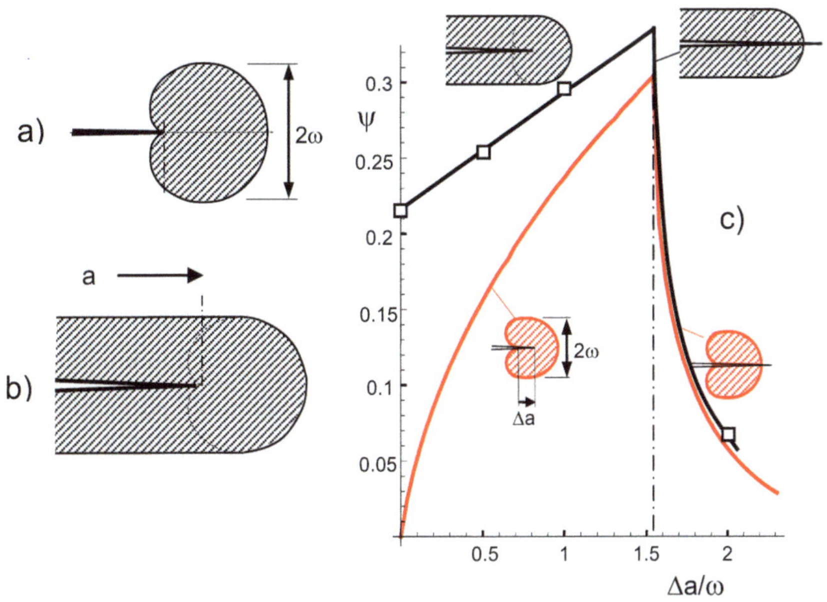

Fig. 7.7 a) Swelling zone for an arrested crack, b) grown crack for weak swelling, c) shielding stress intensity factor for crack extension through the initially swelling zone, red curve: crack passing the initial zone developing for an arrested crack, obtained by application of the procedure by McMeeking and Evans [7.10]; black: grown crack after additional extension Δa.

References

7.1 A. Proctor, I. Whitney, J.W. Johnson, The strength of fused silica, Proceedings of the Royal Society of London. Series A. Mathematical and Physical Sciences, Vol. 297, No. 1451 (1967), 534-557.

7.2 P.J. Lezzi, Q.R. Xiao, M. Tomozawa, T.A. Blanchet, C.R. Kurkjian, Strength increase of silica glass fibers by surface stress relaxation: A new mechanical strengthening method, J. of Non-Crystalline Solids **379** (2013) 95–106.

7.3 S. M. Wiederhorn, G. Rizzi, S. Wagner, M. J. Hoffmann, and T. Fett, Stress-Enhanced Swelling of Silica: Effect on Strength, J. Am. Ceram. Soc. **99**(2016), 2956-63.

7.4 Lawless, J.F. (1974): Statistical Models and Methods for Lifetime Data, Wiley, New York.

7.5 Davis, K.M., Tomozawa, M., Water diffusion into silica glass: structural changes in silica glass and their effect on water solubility and diffusivity, J. Non-Cryst. Sol. **185**(1995), 203-220.

7.6 A. Agarwal, M. Tomozawa, W. A. Lanford, Effect of stress on water diffusion in silica glass at various temperatures, J. Non-Cryst. **167**(1994), 139-148.

7.7 A. Zouine, O. Dersch, G. Walter and F. Rauch, "Diffusivity and solubility of water in silica glass in the temperature range 23-200°C," *Phys. Chem. Glass: Eur. J. Glass Sci and Tech. Pt. B*, **48** [2] 85-91 (2007).

7.8 S.D. Hamann, Chemical equilibria in condensed systems, In: High Pressure Physics and Chemistry, Ed. R.S. Bradley, p. 131, 1963, Academic Press, London.

7.9 S.M. Wiederhorn, "Fracture Surface Energy of Glass," *J. Am. Ceram. Soc.*, **52**[2], 99-105, (1969).

7.10 McMeeking, R.M., Evans, A.G., Mechanics of Transformation-Toughening in Brittle Materials, J. Am. Ceram. Soc. **65**(1982), 242-246.

7.11 G. Rizzi, T. Fett, Defects at Silica Surfaces and crack-tip shielding- A Finite Element Study, Scientific Working Papers **33**, 2015, ISSN: 2194-1629, Karlsruhe, KIT.

7.12 F. Lechenault, D.L. Rountree, F. Cousin, J.-P Bouchaud, L. Ponson and E. Bouchaud, "Evidence of Deep Water Penetration in Silica during Stress Corrosion Fracture," Phys. Rev. Let. **106**(2011), 165504.

8 Prediction of residual deformation

8.1 Swelling stress

In Section 7 and [8.1] we predicted the strength increase of silica fibers by swelling and compared it with measurements by Lezzi et al. [8.2]. Good agreement was found by using the results from Agarwal et al. [8.3] obtained with the Nuclear Resonance Analysis (NRA). In this Section, we compare the swelling effect on residual bending displacements with deformation measurements by Tomozawa et al. [8.4].

In Section 6.3, we derived an equation for the hydroxyl concentration S as a function of the applied stress, σ_{appl}, eq.(6.3.7a)

$$S = -\frac{RT}{\lambda\beta}\mathrm{PLog}\left[-\frac{\lambda\beta S_0}{RT}\exp\left(\sigma_{appl}\frac{\lambda}{RT}\right)\right] \tag{8.1.1}$$

with λ=14.4 cm^3/mol and the parameter β=−135 GPa. The swelling stresses in axial direction are given by eq.(6.3.9) as

$$\sigma_{sw,z} \cong -133 \times S \quad \text{(GPa)} \tag{8.1.2}$$

In this section, the bending moments and curvatures caused by these swelling stresses shall be computed and compared with measurements from [8.4].

Knowledge of the parameter λ in the compression zone of a bent fiber is hardly critical for computations. In compression it principally holds $0 \le S \le S_0$ i.e. allowing only a small region of variation. On the other hand, the stress-affected zone in compression is clearly smaller than in tension due to the stress-affected diffusivity. Consequently, the value of the parameter λ in the pressure range may therefore be approximated to be equal to the tensile case.

8.2 Effect of swelling zones on bending deformation

Volume expansion under tensile loading must have consequences on the deformation behaviour, similar to the strength behaviour discussed in [8.1]. Under bending load, the effect of stress on the SiO_2/water reaction (1.1.1) is stronger in the tensile region than in the compression region. The enhanced swelling under tension and suppressed swelling under compression must result in an internal bending moment acting against the applied one.

In order to give a transparent analysis, let us consider a thin bent fiber of radius r_0, Fig. 8.1a, loaded by a constant displacement between the fiber ends, Fig. 8.1b, and constant force F, Fig. 8.1c. This fiber is water vapour soaked under the externally applied outer fiber bending stress σ_b.

As long as $b<<r_0$ is fulfilled, the bending stress in the diffusion zones is sufficiently constant. The applied stress at the surface as a function of the angle φ is given by

$$\sigma_{appl} = \sigma_b \cos\varphi \tag{8.2.1}$$

The thickness b of the swelling layer is governed by diffusivity and time via

$$b = \sqrt{Dt} \tag{8.2.2}$$

The diffusivity to be used in eq.(8.2.2) is a function of stresses, commonly expressed by the hydrostatic stress component σ_h. When $D(0)=D(\sigma_{appl}=0)$ denotes its value in the absence of an externally applied stress, the diffusivity for the case of stress-enhanced diffusion reads [8.5]

$$D = D(0)\exp\left[\tfrac{1}{3}\sigma_{appl}\frac{\Delta V_w}{RT}\right] \tag{8.2.3}$$

The bending moment by swelling stresses can be written for $b<<r_0$

$$M_{sw} = r_0^2 \int_0^{2\pi} \sigma_{sw}(\varphi)\, b(\varphi) \cos\varphi \, d\varphi \tag{8.2.4}$$

When the swelling zone becomes comparable with the fiber radius, e.g. $b > r_0/5$, the moment has to be computed via

$$M_{sw} = \int_{\varphi=0}^{2\pi} \int_{r=0}^{r_0} \sigma_{sw}(\varphi)(r\cos\varphi)\, r\, dr\, d\varphi \tag{8.2.5}$$

After unloading the internal swelling moment M_{sw} causes a remaining curvature of

$$\frac{1}{R} = -\frac{M_{sw}}{EJ}, \quad J = \frac{\pi}{4} r_0^4 \tag{8.2.6}$$

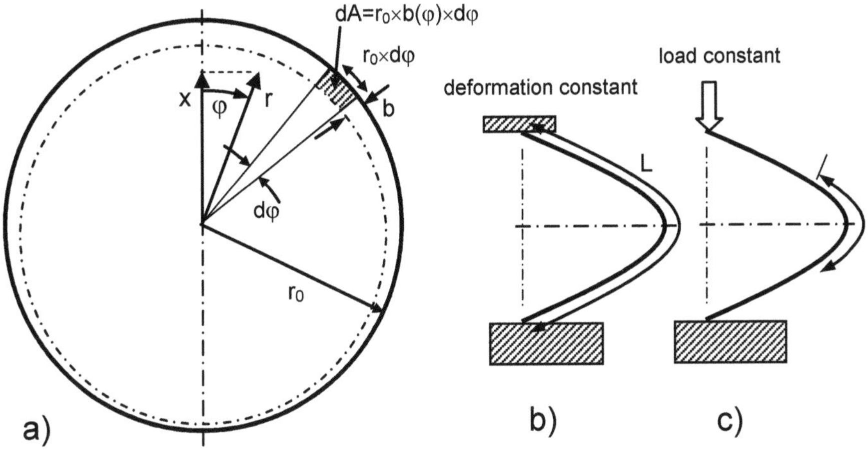

Fig. 8.1 a) Fiber cross-section with swelling zone under bending, b) bending under constant displacements at the ends, c) constant load.

Equation (8.2.4) gives the bending moment by swelling for the case of very thin swelling zones with $b/r_0 \rightarrow 0$. For thicker zones, comparable with the fiber radius r_0, the solution becomes more complicated since then also the equilibrium conditions have to be fulfilled. Because of the Bernoulli hypothesis that plane cross sections must remain plane, the fiber curvature caused by swelling results in a linear strain distribution over the cross section:

$$\varepsilon_z = c_1 x + c_2, \quad x = r\cos\varphi \tag{8.2.7}$$

(for x see Fig. 8.1a).

The integrals of swelling stresses over the cross section of the fiber allow consideration of different types of tests. The requirements of disappearing normal force of swelling stresses and disappearing total moment result in two conditions from which the unknown coefficients c_1 and c_2 can be determined by integrating over the whole cross-section area A

$$\int_{(A)} (\tfrac{1}{E}\sigma_{sw} + c_1 x + c_2)\, dA' = 0 \tag{8.2.8}$$

$$\int_{A} (\tfrac{1}{E}\sigma_{sw} + c_1 x + c_2) x\, dA' = 0 \tag{8.2.9}$$

For constant load, the equations (8.2.8) and (8.2.9) have to be solved simultaneously. For this purpose we used the *Mathematica* [8.6] subroutine *FindRoot*. Boundary conditions for constant displacement are realized by setting c_1=0. Consequently, eq.(8.1.1) now reads

$$S = -\frac{RT}{\lambda\beta}\mathrm{PLog}\left[-\frac{\lambda\beta S_0}{RT}\exp\left(\frac{(\sigma_{appl} + Ec_2 + Ec_1 r\cos\varphi)\lambda}{RT}\right)\right] \tag{8.2.10}$$

8.3 Predictions on the basis of diffusivity affected by applied stresses

8.3.1 Results at 550°C

Figure 8.2a shows the hydroxyl concentration at a water vapour pressure of 355 Torr as a function of temperature as reported by Davis and Tomozawa [8.7]. The data from IR-analyses are indicated by the circles. The ordinate shows the hydroxyl concentration normalized on the value at 650°C. At 550°C, we can conclude that the ratio $S_{550°C}/S_{650°C}$ is about 1.4. The values of S_0 expected from the NR-Analysis [8.3, 8.1] are compiled in Table 8.1 for temperatures ≥550°C. The last column of Table 8.1 gives the diffusivities used for predictions according to eq.(8.2.3). The effect of the different loading conditions is shown in Fig. 8.3a. Since in 2-point-bending tests the bending moment changes along the fiber length axis and due to the fact that the water reaction only took place in a central region (indicated by the arrows in Fig. 8.1c), no well-defined mechanical boundary conditions are given. As limit cases, we consider constant moment and constant curvature.

		NRA	NRA	$D(0)$
Temperature	$S/S_{650°C}$	$S(0)$ (wt%)	S_0 (wt%)	(cm²/s)
650°C	1	0.586	0.652	
550°C	1.4	0.821	0.971	1×10^{-11}
625°C	1.07	0.63	0.71	3×10^{-11}
700°C	0.87	0.51	0.56	7×10^{-11}

Table 8.1 Hydroxyl concentrations from NRA-measurements used for the deformation predictions according to [8.3, 8.1, 8.7].

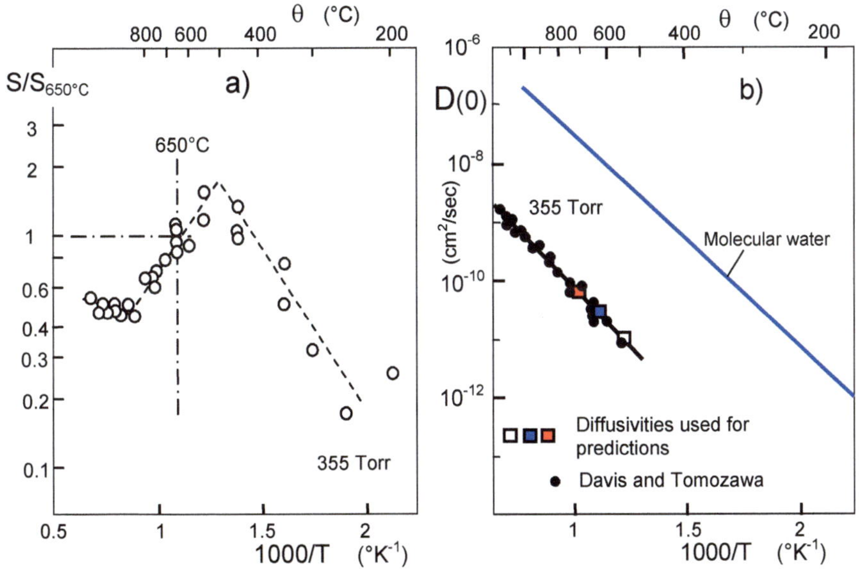

Fig. 8.2 a) Hydroxyl surface concentration S in silica, normalized on the concentration at reference temperature 650°C, $S_{650°C}$ for a water vapour pressure of p= 355 Torr (circles) (Davis and Tomozawa [8.7]); b) Diffusivity from [8.7] (circles); Squares: $D(0)$ used for predictions with eq.(8.2.3).

Computations were performed for a bending stress of 493 MPa at 355 Torr vapour pressure and a diffusivity of $D(0)=1\times10^{-15}$ m²/s as indicated by the white square in Fig.

8.2b. Under constant load ($c_1 \neq 0$, $c_2 \neq 0$) the dash-dotted line resulted and for constant displacement ($c_1=0$, $c_2 \neq 0$), the solid curve was obtained (see Fig. 8.3). The circles represent the experimental data from [8.4]. The effect of the applied bending stress is visible from Fig. 8.3b. The results agree sufficiently with the experimental data from [8.4] for all applied bending stresses. The perpendicular lines indicate an upper limit for the time range in which the swelling layer thickness is smaller than b=10µm. It is self-evident that the residual curvatures cannot increase unlimited with time. For long times the diffused water has homogeneously soaked the full cross section, i.e. S_0=const. The saturation moment results then from eq.(8.2.5).

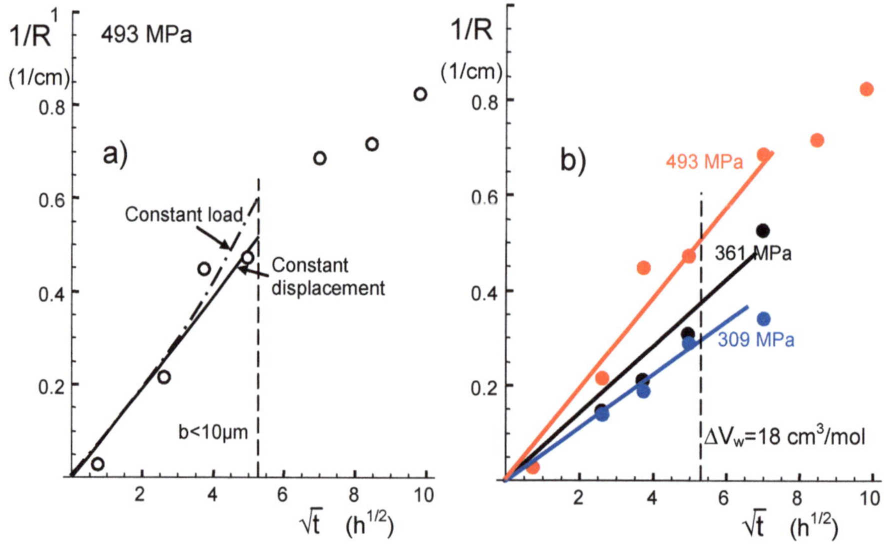

Fig. 8.3 Residual curvature of water vapour soaked silica fibers (125µm diameter, 550°C), computed for $\Delta V_w = 18$ cm³/mol. Circles: results for constant displacement tests by Tomozawa et al. [8.4], a) Effect of different boundary conditions: Solid curve: constant displacement (Fig. 8.1b), dash-dotted curve: constant load (Fig. 8.1c); b) predictions for different loads and constant displacement conditions.

8.3.2 Results at 625 and 700°C

Residual deformations are also available at increased temperatures [8.4]. In addition to the results at 550°C (circles), Fig. 8.4 represents the data for 625°C (squares) and 700°C (triangles), all measured at 493 MPa. The diffusivities for computations are about $D(0)$=3

10^{-15}m^2/s at 625°C (blue square in Fig. 8.2b) and $D(0)=7\times10^{-15}$ m^2/s at 700°C (red square in Fig. 8.2b). The hydroxyl concentration at 625°C is about 7% higher than at 650°C and about 15% smaller at 700°C. For $D(0)$ and S_0 see Table 8.1.

The solid lines in Fig. 8.4a represent the additional computations. Whereas for 550 a tenable agreement between measurement and computations is visible, at higher temperatures of 625 and 700°C the measurements are clearly underestimated by the computations. At 625°C the predictions are 80% and at 700°C about 55% of the measured values. Under the assumption of the diffusivity $D(0)$ as a free parameter, least-squares fits enable to determine the best temperature-dependent values. The resulting diffusivities are entered in Fig. 8.4b by the squares. The related straight-line in Fig. 8.4b is steeper than that for the results by Davis and Tomozawa [8.7].

Whereas for the data of [8.7] the activation is 81 kJ/mol, the value for the squares in Fig. 8.4.b is obtained to be 122 kJ/mol. However, there is also the possibility that the glasses from references [8.4] and [8.7] may behave differently.

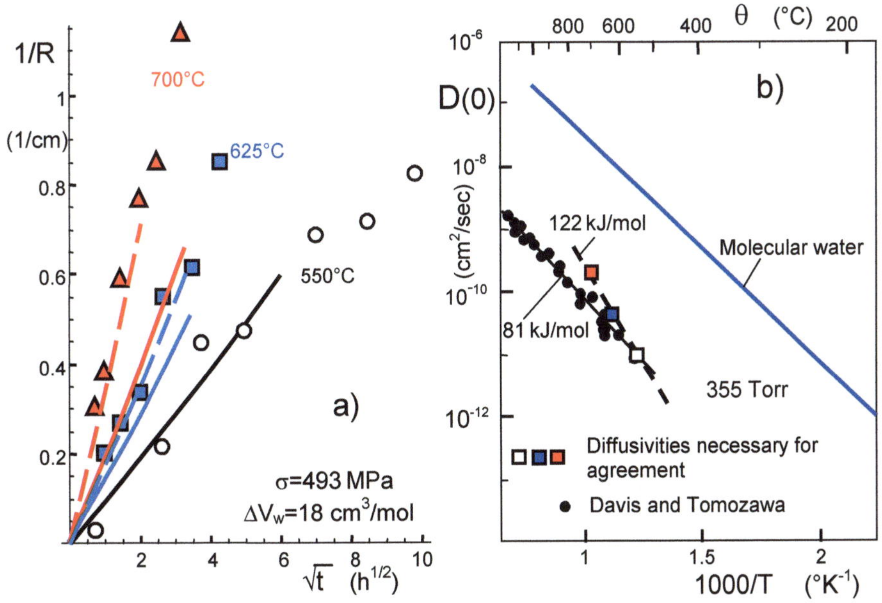

Fig. 8.4 a) Residual curvatures at different temperatures (NRA-data used). Symbols: Measurements by Tomozawa et al. [8.4]. Solid lines: predictions using the diffusivities $D(0)$ from Fig. 8.2b, (Table 8.1), dashed line: Least-squares fit to the measurements. In the case of 550°C the solid and dashed lines are identical; b) diffusivities obtained for σ_{appl}=493 MPa (squares) via data fit in a).

8.4 Predictions including the effect of swelling stresses on diffusivity

The predictions made in Section 8.3 were performed with eq.(8.2.3) that implicitly assumes that the stresses acting in a volume element would be externally applied stresses exclusively. In presence of swelling stresses, this equation must read

$$D = D_0 \exp\left[(\sigma_{h,appl} + \sigma_{h,sw})\frac{\Delta V_w}{RT}\right] \tag{8.4.1}$$

The hydrostatic swelling stress term for *anisotropic swelling*, reads according to Section 6.3

$$\sigma_{h,sw} = \tfrac{1}{3}(\sigma_{sw,z} + \sigma_{sw,y}) = -\frac{\kappa E}{6(1-\nu)}(1+\alpha)S \cong -41\,\text{GPa} \times S \tag{8.4.2}$$

and for isotropic swelling with α=1/3 to be applied in the computation of D_0 from $D(0)$ for the unloaded case σ_{appl}=0,

$$\sigma_{h,sw} = -\frac{2\kappa E}{9(1-\nu)}S \cong -18.7\,\text{GPa} \times S \tag{8.4.3}$$

Equations (8.4.1) and (8.4.3) imply for the unknown diffusivity D_0 in the absence of any stress:

$$D_0 = D(0)\exp\left[18.7\,(\text{GPa}) \times S_0 \frac{\Delta V_w}{RT}\right] \tag{8.4.4}$$

We varied the volume ΔV_w and computed the residual curvatures for the three stresses at 550°C and compared the results with the measurements by using the *Mathematica* Subroutine NonlinearFit [8.6]. The best parameter set was found by matching the computed curves to the measured data by Tomozawa et al. [8.4] as ΔV_w=27 cm^3/mol.

The related set of curves is shown in Fig. 8.5a. Application to the higher temperatures again underestimates the measurements as is visible from Fig. 8.5b. It therefore remains to be stated that a moderate variation in ΔV_w has no significant influence on the predictions.

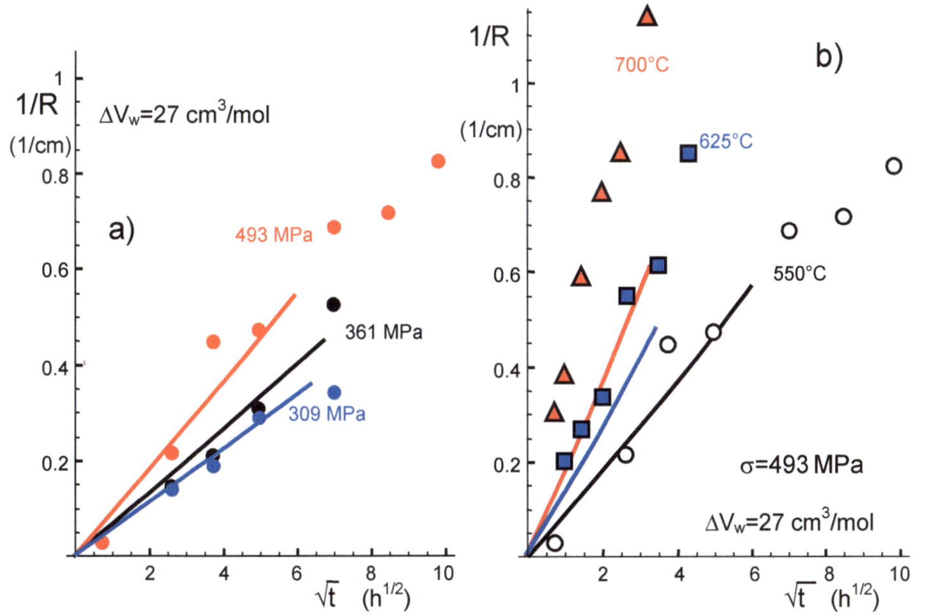

Fig. 8.5 a) Curve fitting according to (8.4.1) for θ=550°C, resulting in ΔV_w=27 cm³/mol, b) predictions for temperatures of 625 °C and 700°C.

8.5 Recovery of curvature after unloading

The hydroxyl concentration at the end of a bending test (σ_{bend} = 493 MPa, t= 36 h) is shown in Fig. 8.6a as the red curve. The S-distribution is non-symmetric with respect to the centreline. On the tensile side, the concentration S is enhanced due to the tensile stresses and on the compression side reduced. The hydroxyl profile in the absence of externally applied stresses is represented by the black curve. This curve is of course symmetric.

If the bent fiber is unloaded and cooled down to room temperature, the S-distribution remains unchanged, and the S-profile caused by bending is "frozen in". No time-dependent "recovery" of the residual curvature is possible.

When the fiber is unloaded at the soaking temperature, the equilibrium constant must decrease because it is no longer stress-enhanced. The consequence is that the equilibrium of reaction (1.1.1) tends more to the left side generating molecular water. This is indicated by the downward arrow in the former tensile region. The time necessary for the reaction can be rather short since diffusion processes are not necessary. In the compression region, the equilibrium constant must increase and, consequently, the *S*-concentration (indicated by an arrow upwards). This reaction is expected to be slow since the additional water must diffuse from the surface.

The changes in *S*-concentration cause changes in volumetric strain as can be directly concluded from eq.(2.1.11). Changes in swelling strains give rise for a change in swelling stresses. In the former tensile region, the swelling stresses decrease and in the former compression region, swelling stresses increase.

The mechanical response is a time–dependent reduction of the curvature (Fig. 8.6b). This type of "relaxation" is of course only a consequence of variation in *S* with time as has to be concluded from the sequence of proportionalities

$$S = f(t) \Rightarrow \sigma_{sw}(t) \Rightarrow M_{sw}(t) \Rightarrow 1/R(t) \tag{8.5.1}$$

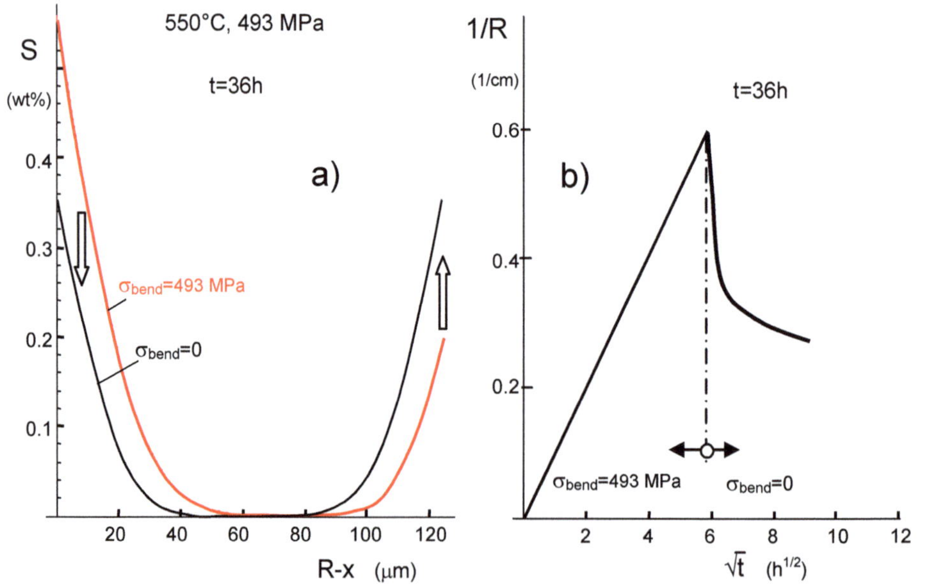

Fig. 8.6 a) Hydroxyl concentration along the diameter of a bent fiber (red curve) compared with the *S* profile in the absence of bending, b) expected mechanical response of curvature for changes in the equilibrium constant and hydroxyl concentration.

References

8.1 S. M. Wiederhorn, G. Rizzi, S. Wagner, M. J. Hoffmann, and T. Fett, Stress-Enhanced Swelling of Silica: Effect on Strength, J. Am. Ceram. Soc. **99**(2016), 2956-63.

8.2 P.J. Lezzi, Q.R. Xiao, M. Tomozawa, T.A. Blanchet, C.R. Kurkjian, Strength increase of silica glass fibers by surface stress relaxation: A new mechanical strengthening method, J. of Non-Crystalline Solids **379** (2013) 95–106.

8.3 A. Agarwal, M. Tomozawa, W. A. Lanford, Effect of stress on water diffusion in silica glass at various temperatures, J. Non-Cryst. **167**(1994), 139-148.

8.4 M. Tomozawa, P.J. Lezzi, R.W. Hepburn, T. A. Blanchet, D. Cherniak, Surface Stress Relaxation and Resulting Residual Stress in Glass Fibers: A New Mechanical Strengthening Mechanism of Glasses, J. Non-Cryst. Solids **358**(2012), 2650.

8.5 P.G. Shewmon, Diffusion in Solids, McGraw-Hill, New York, 1963.

8.6 *Mathematica*, Wolfram Research, Champaign, USA.

8.7 Davis, K.M., Tomozawa, M., Water diffusion into silica glass: structural changes in silica glass and their effect on water solubility and diffusivity, J. Non-Cryst. Sol. **185**(1995), 203-220.

9 Comparison between prediction and experiment

9.1 Comparison of experimental results with predictions from swelling

In this Section the experimental and predicted results that scattered over the previous subject areas are summarized and compared. These results are compiled in Table 9.1 and Fig. 9.1 sequential numbers in the first column of the table are intended to facilitate the assignment of the data in the tables to those in the text and in the figure.

No.	Property	predicted	experimental	ratio
(1)	$-M_b$ (90°C) (Nm/m) [9.1]	0.0047	0.00413 [0.00363, 0.00463]	0.878 [0.772, 0.985]
(2)	$-M_b$ (90°C) (Nm/m) [9.1]	0.0040		1.032 [0.908, 1.16]
(3)	$-M_b$ (200°C) (Nm/m) [9.1]	0.0145	0.0173 [0.0128, 0.0217]	1.19 [0.882, 1.496]

Table 9.1 Comparison of predicted properties by using swelling effect and measured data.

Under (1) the bending moment for water-soaked silica disks at 90°C by Wiederhorn et al. [9.1] is given together with the prediction based on water content via IR-measurement (see Fig. 2.3). The experimental numbers in brackets represent ±1 standard deviation (S.D.).

Under (2) the same experimental data are compared with the predictions using water concentrations at the surface as reported by Zouine et al. [9.2].

Finally, (3) shows the results for the tests at 200°C (see Fig. 2.5). In Fig. 9.1 all these data are plotted as the ratio of measured results and predictions. It can be seen that these ratios do not significantly deviate from the value "1", i.e. predictions and measurements are in rather good agreement. In Sections 7 and 8, the predictions of strength increase by swelling stresses [9.3] and residual deformation in 2-point bending tests were made based on the value $V_0\varepsilon^{(1)}_{sw,z} = 14.4$ cm^3/mole as was concluded in [9.3].

The strength predictions are compiled in Fig. 9.2a by the solid curves (based on NRA- and IR-measurements) together with the data from [9.4]. In Table 9.2, the measured and predicted strength increases are compiled. Numbers in brackets represent ±1 Standard Deviation. In addition, these results are introduced in Fig. 9.1.

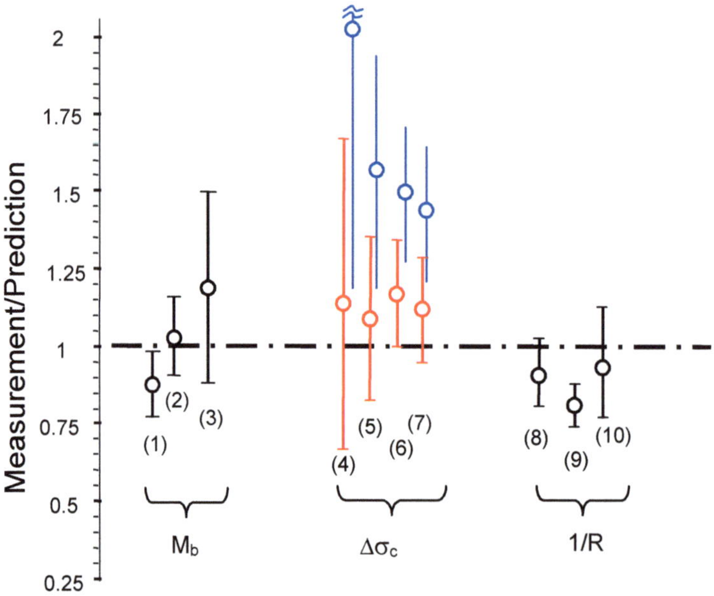

Fig. 9.1 Comparison of predicted properties by using swelling effect and measured data. The serial numbers of the tests are given in parentheses, see Tables 9.1-9.3.

For the second data set in Fig. 9.1, where strength increases are considered, the fact that about 10 % of the 15 test specimens did already fail under the 3 GPa pre-stressing has an influence on the strength of the surviving specimens as has been outlined in Section 7.1. The slightly reduced strength at 3 GPa is shown in Fig. 9.3 and introduced as test (7) in

Fig. 9.1 and Table 9.2. In both cases, the predictions via the NRA-concentrations are introduced as red circles, bars, and numbers. However, if the predictions are based on the IR data, there are larger deviations (Table 9.2 and Fig. 9.2). The predictions based on the IR-data are given in blue colour.

The residual curvatures after 2-point bending at 550°C [9.5] are shown in Fig. 9.2b. In Table 9.3, the measured and predicted residual curvatures are compiled. The numbers in brackets represent the 90% Confidence Intervals.

		Strength increase $\Delta\sigma_c$		
		measured	predicted	measured/predicted
No.	σ (GPa)			
(4)	1	0.51 [0.297, 0.754]	0.45 0.25	1.13 [0.66, 1.67] 2.04 [1.19, 3.0]
(5)	2	1.21 [0.92, 1.49]	1.11 0.77	1.095 [0.83, 1.35] 1.57 [1.19, 1.94]
(6)	3	2.205 [1.89, 2.54]	1.89 1.48	1.17 [1.00, 1.34] 1.49 [1.28, 1.72]
(7)	3	2.11 [1.80, 2.44]	1.89 1.48	1.12 [0.95, 1.29] 1.43 [1.21, 1.65]

Table 9.2 Measured [9.4] and predicted strength increase $\Delta\sigma_c$ [9.3], red: via NRA, blue: via IR.

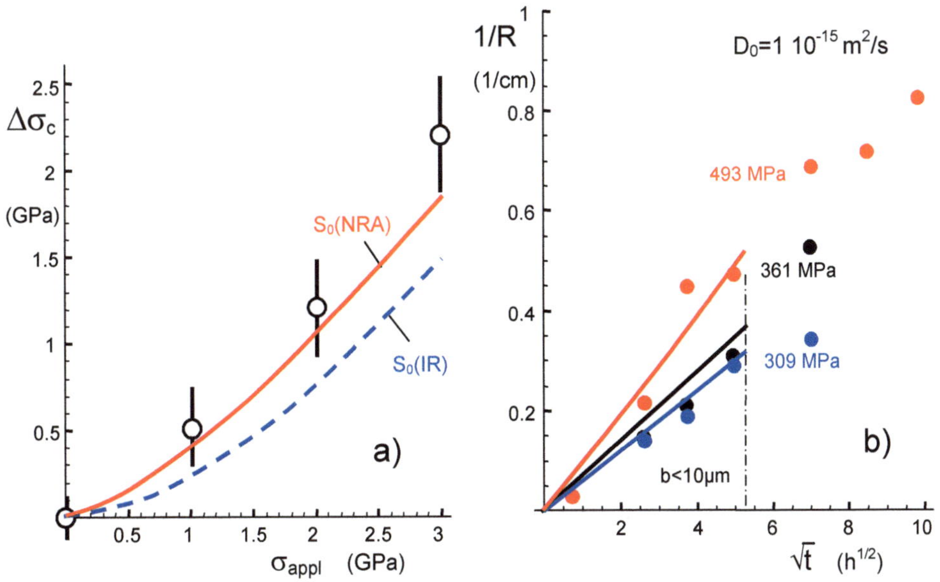

Fig. 9.2 Predictions compared with measurements (red: S_0 from NRA, blue: from IR-analysis), a) strength predictions, b) residual curvature for 550°C.

		Residual curvature $(1/R)/\sqrt{t}$		
No.	σ (MPa)	measured	predicted	measured/predicted
(8)	309	0.056 [0.050, 0.063]	0.062	0.903 [0.806, 1.02]
(9)	361	0.060 [0.055, 0.065]	0.074	0.81 [0.74, 0.88]
(10)	493	0.101 [0.08197, 0.122]	0.108	0.935 [0.76, 1.13]

Table 9.3 Measured and predicted residual curvature (curve slopes: R in cm, t in h).

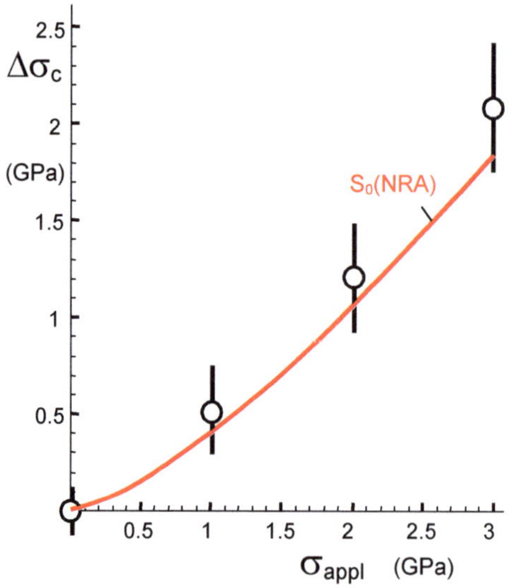

Fig. 9.3 Strength predictions compared with measurements (data point at 3 GPa corrected according to Fig. 7.2b).

Concluding remark

In our opinion, the mechanical effect of the water/silica reaction is a volume increase, i.e. a swelling of the hydroxyl containing regions. This effect was experimentally proved via density measurements by Brückner [9.6, 9.7], Shackelford [9.8] and Shelby [9.9]. As illustrated in Fig. 9.1 there is good agreement between predictions on swelling and experimental results. This holds also for the strength and residual curvatures in Fig. 9.2. In this context, it has to be taken into account that the strength predictions only included the global swelling effect and ignored strength-increasing local swelling around the individual surface cracks.

Our description is based on equation (2.1.11) relating the hydroxyl concentration S as the product of the silica-water-reaction (1.1.1) to a volume increase ε_v. The proportionality factor has to be understood as a material-specific property.

Time dependent effects must of course occur during experiments. This is the case when the hydroxyl concentration depends on time:

1. In soaking experiments there is first an increase of the molecular water at the surface (see Section 1) with the consequence $C(t)$, $S(t)$.

2. For a fixed given volume element below the surface there is in addition a continuously increase in water due to the diffusion process.

3. Finally, the equilibrium is not spontaneously reached but may be approached with time delay due to finite reaction rates especially at lower temperatures.

In all these cases the reason for mechanical responses is volume swelling and any time effect is caused by varying S-concentrations

$$\varepsilon_v(t) = \kappa \times S(t) \qquad (9.1.1)$$

For an application see Section 8.5, where recovery effects are discussed which have to be expected after bending tests when the temperature is not reduced.

9.2 Comparison with other explanatory approach

Some studies focus on stress relaxation as a possible explanation for the effects of the silica/water reaction. But effects occurring in the absence of pre-existing stresses as for instance the measured generation of bending moments in completely stress-free silica disks cannot be a consequence of any type of relaxation.

Our *conclusions* from the point of view of the disk results [9.1] are:

- Before the tests, the silica disks were etched in 5% volume fraction HF in order to remove 5-10μm of glass, which might have contained a diffusion layer of water at the surface or might have residual stresses due to polishing. Consequently, stress relaxation could not be the reason for curvature since no stresses to relax were present.

- We have to conclude that the mechanical effect of the reaction (1.1.1) is an *active process* that generates strains and stresses depending on the mechanical boundary conditions. It is not a *passive* effect as is stress relaxation that needs pre-existing stresses.

- On the other hand, Fig. 9.1 clearly shows that the predictions and measurements of curvature (or bending moments) are within 0.85-1.20 of the ideal agreement (dash-dotted line =1) for the case that the NRA results are used. We have to state that there is *no free space* for any further stress-generating effect. This confirms the assumption that the water species causing swelling strains is exclusively the hydroxyl S even at rather low temperatures.

- Moreover that it can be concluded that no relaxation is present at 200°C. If relaxation would be present at 200°C, the generated stresses causing the observed curvature must have relaxed simultaneously to their creation and of course not any remaining curvature could result in contradiction to the observation.

Conclusions with respect to time-dependent diffusion profiles and diffusivity, Sections 1 and 3:

- It could be shown that the change of the diffusion profile shapes with increasing time is a trivial consequence of the penetration limiting *reaction parameter* or *mass transfer coefficient* in combination with swelling stresses.

- The same holds for the apparent time-dependence of the diffusivity. An interpretation in terms of relaxation effects is to our knowledge not available.

It should be emphasized again that the effect of swelling is already proved experimentally by the density reduction due to hydroxyl generation [9.6-9.9]. Direct measurements of a relaxation are not known after knowledge of the authors up to now.

Conclusion with respect to the strength and residual curvature results of silica fibers:

- The explanation of the results in Fig. 9.2, given in [9.4, 9.5], is based on a complete relaxation of pre-existing bending and tensile stresses in the thin surface layer containing water. After unloading, these layers should show an inverse stress. Unfortunately, the authors do not give a critical water concentration, at which the total stress relaxation should take place.

- Our competing explanation of the mechanical effect of reaction (1.1.1), suggested in [9.10, 9.11, 9.12], is local swelling of the glass volume due to hydroxyl generation as was experimentally proved in literature [9.6-9.9]. By mechanical boundary conditions, the swelling strains result in swelling stresses proportional to the amount of hydroxyl concentration. This attempt is appropriate to predict quantitatively the mechanical effect of swelling on strength and residual curvature [9.4, 9.5] and allows a qualitatively understanding of effects by simplified computations as were presented here (see Sections 6-8).

- If relaxation exists, this effect is most likely to manifest itself in the tensile strength measurements. Table 9.2 makes this clear under the assumption that the IR data must be used for prediction. This could explain the remaining discrepancies between the prediction via the swelling effect and the measurement results.

From the standpoint of our current investigations, we can interpret many of the experimentally available results in terms of volume swelling. Nevertheless, we consider relaxation and also an interaction of swelling and relaxation to be quite conceivable. Swelling occurs under tension and pressure (to a lesser extent) and, in our opinion, a catalytic effect of water molecules could occur only in the tensile region but not under compressive stress.

A possible reason for the existence of an apparent relaxation effect exclusively under tensile stress is addressed in Appendix A4. From modelling the breaking of SiO_2 bonds by water molecules catalytically without involving a chemical reaction according to eq. (1.1.1) the possibility of a stress reduction appears to be simply possible.
This was suggested very early on by the authors in a purely mechanical model analysis [A4.13, A4.14, A4.15].

In addition to the power law of subcritical crack growth, $v \propto K^n$, this model provided two main results:

- The occurrence of a threshold value for permanent bond fracture by molecular water.

- On the other hand, it was implicitly found that only <u>tensile</u> stresses in the glass show this effect.

- Furthermore, a threshold of the applied load should occur.

These results can be transferred to the strength behavior under the influence of molecular water by replacing the stress intensity factors at the crack with stresses in the glass. Especially a *stress threshold* value should be of importance for the application of this model on strength behavior of silica.

The question that has to remain unanswered at the moment is whether the bond splitting will completely or partially decrease again after unloading. This question is irrelevant in the case of subcritical crack growth, since after each break of a bond the tip of the crack advances to the next bond and very large crack openings occur at their original location, which ruled out re-bonding.

References

9.1 S. M. Wiederhorn, F. Yi, D. LaVan, T. Fett, M.J. Hoffmann, Volume Expansion caused by Water Penetration into Silica Glass, J. Am. Ceram. Soc. **98** (2015), 78-87.

9.2 A. Zouine, O. Dersch, G. Walter and F. Rauch, "Diffusivity and solubility of water in silica glass in the temperature range 23-200°C," *Phys. Chem. Glass: Eur. J. Glass Sci and Tech. Pt. B*, **48** [2] 85-91 (2007).

9.3 S. M. Wiederhorn, G. Rizzi, S. Wagner, M. J. Hoffmann, and T. Fett, Stress-Enhanced Swelling of Silica: Effect on Strength, J. Am. Ceram. Soc. **99** (2016), 2956-63.

9.4 P.J. Lezzi, Q.R. Xiao, M. Tomozawa, T.A. Blanchet, C.R. Kurkjian, Strength increase of silica glass fibers by surface stress relaxation: A new mechanical strengthening method, J. of Non-Crystalline Solids **379** (2013) 95–106.

9.5 M. Tomozawa, P.J. Lezzi, R.W. Hepburn, T. A. Blanchet, D. Cherniak, Surface Stress Relaxation and Resulting Residual Stress in Glass Fibers: A New Mechanical Strengthening Mechanism of Glasses, J. Non-Cryst. Solids **358** (2012), 2650.

9.6 Brückner, R., "The structure-modifying influence of the hydroxyl content of vitreous silicas," Glastech. Ber. **43**(1970), 8-12.

9.7 Brückner, R., "Metastable equilibrium density of hydroxyl-free synthetic vitreous silica," J. Non-Cryst. Solids, **5**(1971), 281-5

9.8 Shackelford, J.F., Masaryk, J.S., Fulrath, R.M., "Water Content, Fictive Temperature and Density Relations for Fused Silica," J. Am. Ceram. Soc. **53** (1970), 417.

9.9 Shelby, J.E., “Density of vitreous silica,” J. Non-Cryst. **349** (2004), 331-336.

9.10 S.M. Wiederhorn, T. Fett, G. Rizzi, S. Fünfschilling, M.J. Hoffmann and J.-P. Guin, “Effect of Water Penetration on the Strength and Toughness of Silica Glass,” *J. Am. Ceram. Soc.* **94** (2011) [S1] S196-S203.

9.11 S.M. Wiederhorn, T. Fett, G. Rizzi, M. Hoffmann, J.-P. Guin, “Water Penetration – its Effect on the Strength and Toughness of Silica Glass,” *Met. Mater. Trans. A*, **44** (2013) [3], 1164 -1174.

9.12 T. Fett, G. Rizzi, M. Hoffmann, S. Wagner, and S.M. Wiederhorn, “Effect of Water on the inert Strength of Silica Glass: Role of Water Penetration,” *J. Am. Ceram. Soc.* **95** (2012) [12], 3847-3853.

9.13 Fett, T. (1992): Interpretation of subcritical crack growth in ceramic materials, DFVLR-Forschungsbericht, 82-07, 1982, (in German).

9.14 Fett, T., A fracture-mechanical theory of subcritical crack growth in ceramics, Int. J. Fract. **54**(1992), 117–130.

9.15 Munz, D., Fett, T., Ceramics, Failure, Material Selection, Design. Springer, Berlin Heidelberg New York (1999).

10 Dissolution effect for liquid-water soaking

10.1 Dissolution data from literature

Although fused silica is a very corrosion resistant material, it shows at higher temperature dissolution effects in liquid water and, consequently, a surface removal. Diffusion layer thicknesses will be reduced compared to surfaces undergoing water vapour soaking. Consequently, swelling layers will be partially removed.

An estimation of the dissolution effect needs the dissolution rate and the saturation concentration. The thickness of the dissolved surface layer depends on the dissolution rate D_{dis} and the concentration of dissolved SiO_2 in the water, C_{SiO2}, described by the classical power law relation

$$D_{dis} = D_{dis,0}\left(1-\frac{C_{SiO2}}{C_{eq}}\right)^n \tag{10.1.1}$$

with the dissolution rate $D_{dis,0}$ for C_{SiO2}=0 and the equilibrium concentration $C_{SiO2,\infty} = C_{eq}$. The exponent is often given as n>2.

Dissolution rates $D_{dis,0}$ for eq.(10.1.1), obtained from literature are plotted in Fig. 10.1a. The circles correspond to a soda-lime glass that should have the same dissolution behavior as silica [10.1]. The square shows the rate for silica at 150°C in a 0.0167 M $CaCl_2$ solution [10.2]. From these data we expect e.g. for 250°-260°C, where the specimens for the strength measurements in [10.3, 10.4] were prepared, a dissolution rate of $D_{dis} \approx 1$ nm/s (triangle in Fig. 10.1a) in buffered water environment. This high rate implies that long-time soaking in water must cause a strong surface removal. This would of course only be the case if always fresh water is available at the silica surface.

For 192h at 250°C it would result a surface removal of about 550 μm. The pure water in our experiment and the results for the 0.0167 M $CaCl_2$ solution may differ somewhat, nevertheless the dissolution effect would be dramatic.

10.2 Effective diffusion zone size

In order to estimate the maximum effect of dissolution, any redeposition of dissolved material may be ignored in the following computation (i.e. a specimen in a large autoclave is assumed).

The swelling zone size b given as

$$b = \sqrt{D_w t} \tag{10.2.1}$$

increases with respect to time t by a rate of

$$\frac{db}{dt} = \frac{d}{dt}\sqrt{D_w t} = \tfrac{1}{2}\sqrt{\frac{D_w}{t}} = \tfrac{1}{2}\frac{D_w}{b} \tag{10.2.2}$$

If the dissolution rate in water of 250°C is abbreviated as $D_{dis}=\mathrm{d}W/\mathrm{d}t$, with half specimen thickness W, the total zone thickness results as

$$\frac{db_{eff}}{dt} = \tfrac{1}{2}\sqrt{\frac{D_w}{t}} - D_{dis} \tag{10.2.3}$$

A final steady-state zone thickness is approached for $\mathrm{d}b_{eff}/\mathrm{d}t=0$ instead of a continuously increasing thickness as suggested by eq.(10.2.1). This simply results in an equilibrium thickness reached after the time

$$t_{eq} = \tfrac{1}{4}\frac{D_w}{D_{dis}^2} \tag{10.2.4}$$

Equation (10.2.3) reads in integrated form

$$b_{eff} = \begin{cases} \sqrt{D_w t} - D_{dis} t & for \quad t < t_{eq} \\ b_{eq} & \quad t \ge t_{eq} \end{cases} \tag{10.2.5}$$

For equilibrium:

$$b_{eq} = \tfrac{1}{4}\frac{D_w}{D_{dis}} \tag{10.2.6}$$

From D_{dis} = 1 nm/s and $D_w = 6.2\times10^{-16}$ m^2/s at 260°C it results a nearly negligible zone depth of only $b_{eff} \approx 0.16$ µm reached after a few minutes.

10.3 Saturation of the water environment

The rather dramatic rate effect is fortunately limited by the saturation uptake of SiO_2 in water. The dissolution effect in a limited water environment is counteracted by the increase of the silica concentration in the liquid as results from eq.(10.1.1) and the increase of redeposition. The saturation concentration C_{eq} increases with temperature, too. Figure 10.1b shows an average curve that represents literature data for the equilibrium concentration C_{eq}, compiled by Gunnarsson and Arnorsson [10.5].

In our autoclave the water volume was less the glass volume. The solubility at a soaking temperature of for instance 250°C results from Fig. 10.1b as 1250 ppm [10.5], i.e. C_{eq} = 0.00125.

In the case of narrow packed specimens with a water volume smaller than the glass volume as was the case for our soaking procedure in [10.3, 10.4], a simplified estimation of the dissolution can be based on the equilibrium concentration exclusively, resulting in the maximum possible effect.

The maximum dissolved silica in the water containment of volume V_w and density $\rho_w = 1$ g/cm^3 is $m_{dis} = V_w C_{eq}$. The amount of silica removed in a surface layer of area S_{glass}, thickness d and density $\rho = 2.2$ g/cm^3 is $m_{glass} = S_{glass} \times \rho_{glass} \times d$. Since $m_{dis} = m_{glass}$, it follows

$$d = \frac{V_w}{S_{glass}}\frac{\rho_w}{\rho_{glass}}C_{eq} \tag{10.3.1}$$

For our tests we used a very small autoclave of 30 cm^3 total volume. After placing a number of N specimens of each V_{glass}=0.3×0.4×4.5cm^3=0.54 cm^3 the autoclave was filled with water to maximum 2/3 of its height. This corresponds to the water volume

V_w=20cm^3 – $N\times V_{glass}$. The total surface of the N silica specimen is S_{glass}=2N(0.3+0.4) 4.4 cm^2 + 2N×0.3×0.4 cm^2 =N6.4 cm^2.

By introducing the data estimated before into eq.(10.3.1) with the number of N=6-7 as was used for most of the soaking treatments, it results for the maximum possible layer removal: d=2.5μm.

The diffusion layer generated during 24h water soaking is with D=4.6×10^{-16} m^2/s: b= 6.3 μm. Comparison of d and b makes clear that dissolution is not negligible especially for rather short soaking times.

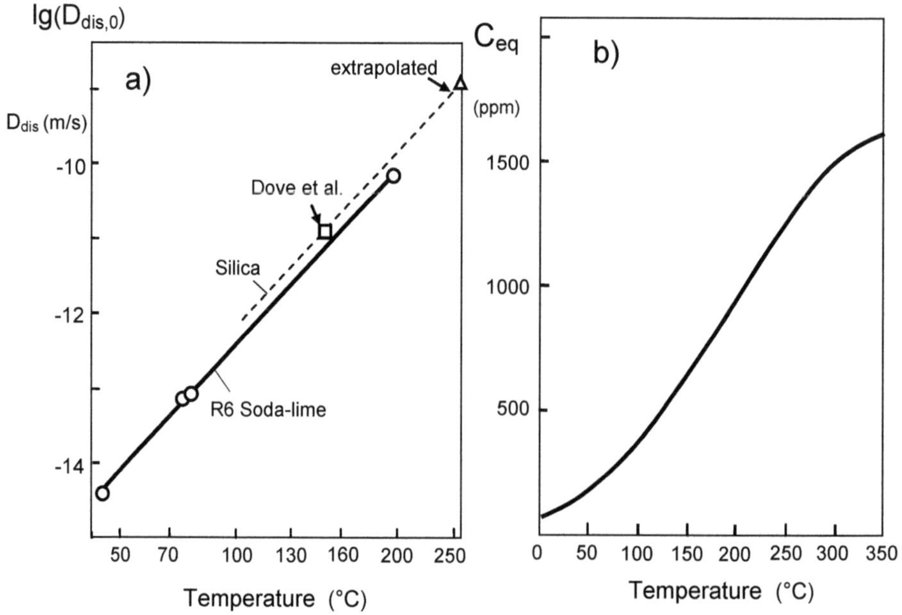

Fig. 10.1 a) Rate for dissolution of silica in water (circles: data for soda-lime glass R6 with about equal dissolution behaviour as silica, reported by Perera and Doremus [10.1], square: result from Dove et al. [10.2], triangle: expected for 260°C), b) solubility of silica in water based on a literature review by Gunnarsson and Arnorsson [10.5].

References

10.1 G. Perera, R.H. Doremus, Dissolution rate of silicate glasses in water at pH 7, J. Am. Ceram. Soc. **74** (1991), 1269-74.

10.2 P.M. Dove, N. Han, A.F. Wallace, J.J. De Yoreo, Kinetics of amorphous silica dissolution and the paradox of the silica polymorphs, PNAS, **105**, [29] (2008), 9903-9908.

10.3 S. Wagner, B. Hohn, D. Creek, M. J. Hoffmann, G. Rizzi, S. M. Wiederhorn, T. Fett, Inert strength measurement on hot water soaked silica, Scientific Working Papers **19**, 2014, ISSN: 2194-1629, Karlsruhe, KIT.

10.4 P. Hettich, S. Wagner, G. Schell, T. Fett, Inert strength measurement on silica soaked at 250°C in liquid water and water vapour, Scientific Working Papers **122**, 2019, ISSN: 2194-1629, Karlsruhe, KIT.

10.5 I. Gunnarsson, S.Arnorsson, Amorphous silica solubility and the thermodynamic properties of H_4SiO_4 in the range of 0° to 350°C at Psat, Geochimica of Cosmochimica Acta, **64** (2000), 2295-2307.

APPENDIX

A1 Crack problems

A1.1 Single edge crack

In the following considerations it is assumed for clearness that the equilibrium of reaction (1.1.1) is already reached. A single crack embedded in a swelling zone of height ω and constant volume swelling is shown in Fig. A1.1. The crack length, *a*, and the zone height are negligibly small compared to the specimen dimensions.

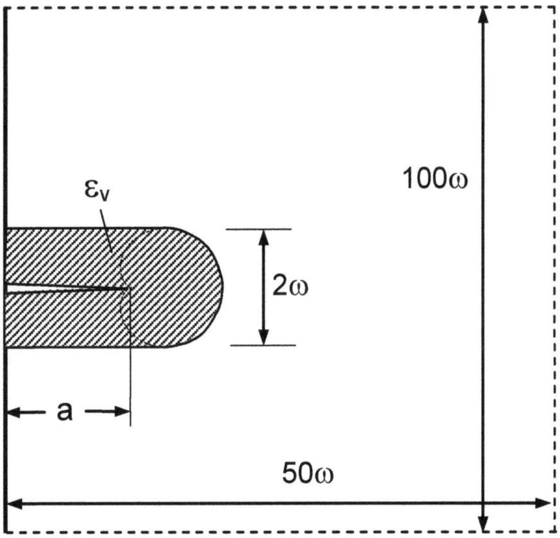

Fig. A1.1 Single edge crack in a semi-infinite specimen embedded in a swelling zone.

Zones of different heights were modelled in a finite element (FE) study. The case of a half-space was realized by a plate of width 50ω and height 100ω. Solid continuum elements (8-node bi-quadratic) were chosen and the computations carried out with ABAQUS Version 6.8. The volume strain was replaced by the equivalent thermal problem by heating the swelling zone by ΔT keeping T=0 in the rest of the structure.
This results in the strain

$$\varepsilon_v = 3\alpha\Delta T \tag{A1.1.1}$$

(α=thermal expansion coefficient).

The shielding stress intensity factors from FE-computations normalized on the value $K_{sh}(0)$, i.e. for the semi-infinite crack with ω/a=0 [A1.1], are plotted in Fig. A1.2 as a function of the ratio of ω/a. The FE-results given by the circles can be expressed by the relation

$$K_{sh}(\omega/a) \cong K_{sh}(0)\exp\left[-1.18\frac{\omega}{a}\right] \tag{A1.1.2}$$

with

$$K_{sh}(0) = -\psi(0)\frac{\varepsilon_v E}{1-\nu}\sqrt{\omega} \tag{A1.1.3}$$

and different zone shape

$$\psi(0) = \begin{cases} 0.22 & \text{for heart shape} \\ 0.25 & \text{for semi circle} \end{cases} \tag{A1.1.4}$$

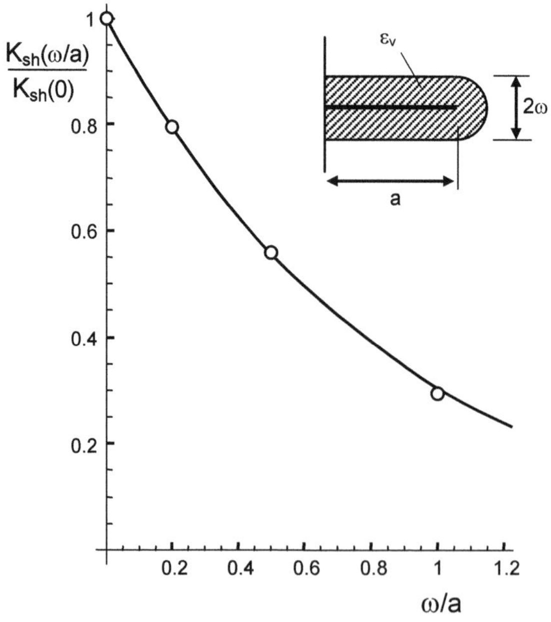

Fig. A1.2 Shielding stress intensity factor as a function of the relative zone height ω/a.

From the solution by eqs.(A1.1.2-A1.1.4), the shielding stress intensity factor for any swelling distribution can be determined using the Green's function as outlined in [A1.2]. The shielding contribution of a zone increment of thickness dω', (Fig. A1.3), is

$$dK_{sh} = -\psi \frac{\varepsilon_v E}{(1-\nu)\sqrt{\omega'}} 1.18 \exp\left[-1.18\frac{\omega'}{a}\right]\left(0.4237 - \frac{\omega'}{a}\right) d\omega' \quad \text{(A1.1.5)}$$

defining the Green's function

$$h(\omega') = -\psi \frac{E}{(1-\nu)\sqrt{\omega'}} 1.18 \exp\left[-1.18\frac{\omega'}{a}\right]\left(0.4237 - \frac{\omega'}{a}\right) \quad \text{(A1.1.6)}$$

This function allows to determine the shielding stress intensity factor for an arbitrarily given strain distribution ε(ω')

$$K_{sh} = \int_0^\infty \varepsilon(\omega')\, h(\omega')\, d\omega' \quad \text{(A1.1.7)}$$

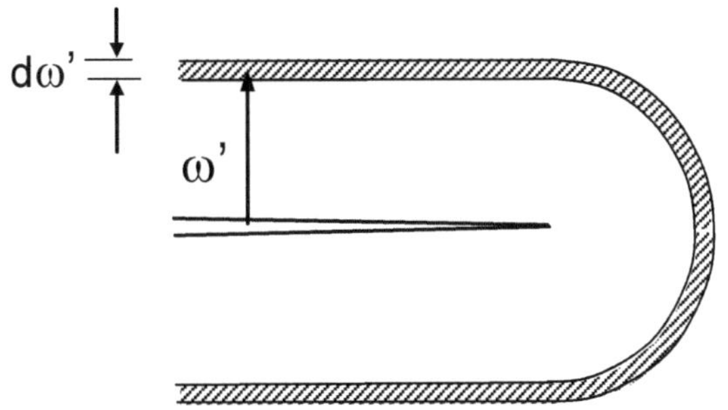

Fig. A1.3 Shape of a differential layer of thickness dω' in a zone with constant hydroxyl concentration.

A1.2 Multiple edge cracks

A1.2.1 Applied stress intensity factor

Fig. A1.4a shows an array of identical edge cracks. The applied stress intensity factors are known from literature [A1.3]. We can represent these data by the simple relation [A1.4]

$$K_{appl} = \sigma_{appl} \sqrt{c/2} \tanh^{5/11} \left(1.122 \sqrt{2\pi} \sqrt{\frac{a}{c}} \right)^{11/5} \tag{A1.2.1}$$

with the *constant* solution for long cracks, $a>c/3$

$$K_{appl} = \sigma_{appl} \sqrt{c/2} \tag{A1.2.2}$$

independent of the crack depth a. This fact is very important. Equation (A1.2.1) is plotted in Fig. A1.4b. For half-penny shaped surface cracks, the solution deviates for $a/c>0.5$ less than 10% as can be seen from Isida and Tsuru [A1.5]. In the following computations we therefore consider the surface cracks as edge cracks.

Whereas the single-crack solution increases continuously with increasing crack length, the stress intensity factor for the crack array remains constant for larger than $a/c \approx 1/3$.

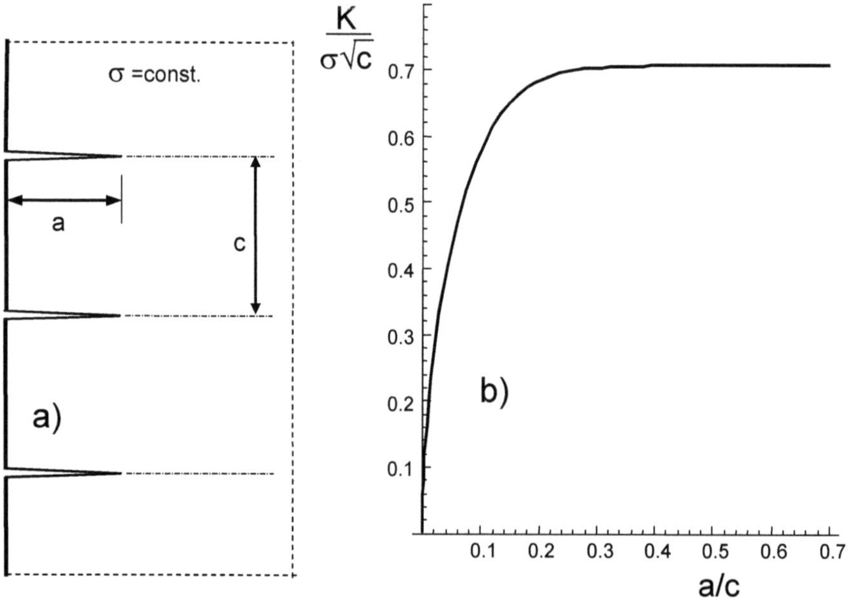

Fig. A1.4 a) Array of edge cracks under remote load σ, b) stress intensity factor solution.

A1.2.2 Shielding stress intensity factor

Swelling behaviour of multiple cracks is the same as for a single crack. This is illustrated in Fig. A1.5a for a subcritically grown crack with heart-shaped zone front contour and in Fig. A1.5b for a crack array showing semi-circular zone fronts in the absence of external load.

The crack array with the semi-circular zone front shape and the same swelling zone of height ω for each crack yields the shielding stress intensity factors of Fig. A1.6.

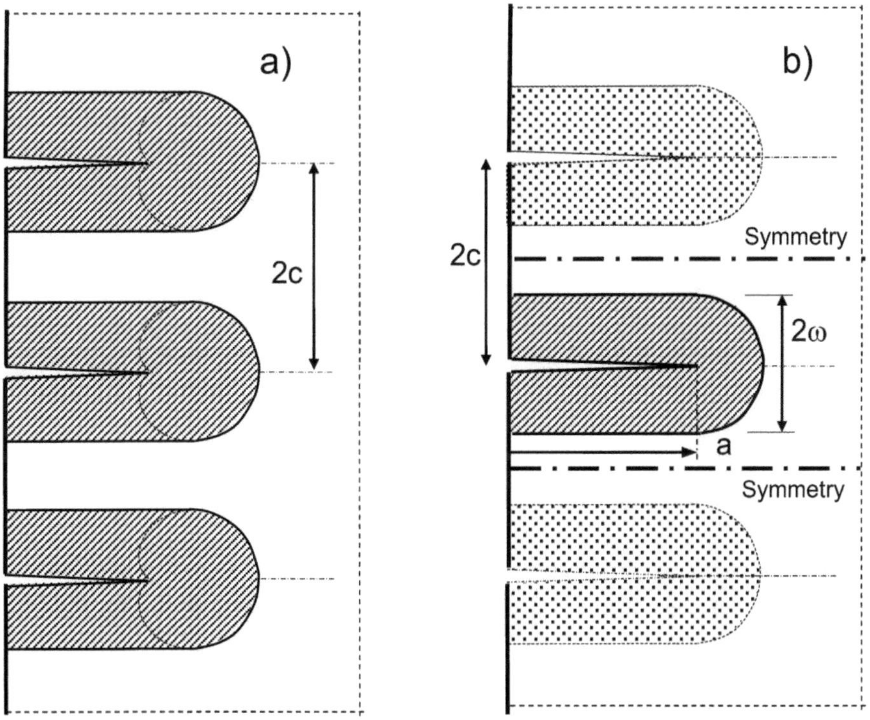

Fig. A1.5 a) Swelling zones of growing multiple cracks, b) modelling by a single crack introducing symmetry planes.

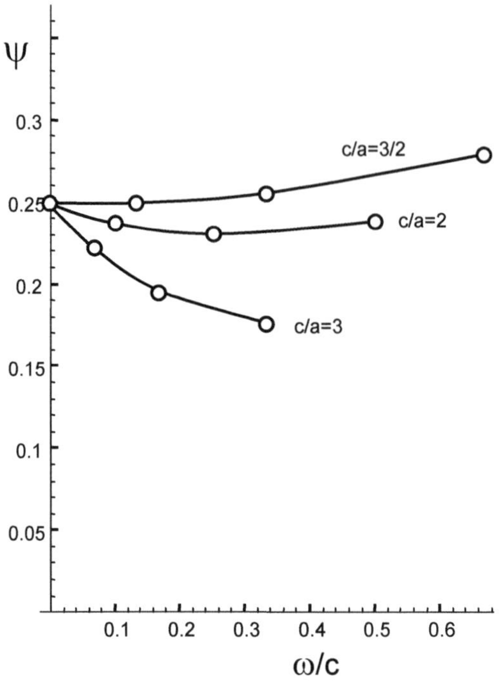

Fig. A1.6 Shielding stress intensity factors for multiple cracks (semi-circular zone ends).

A1.3 Stress-enhanced swelling zones at short surface cracks

In the case of a *long crack* of length a with a swelling zone height ω small compared to the crack length, $\omega \ll a$, the heart-shaped zone at a crack tip yields a disappearing shielding stress intensity factor $K_{sh}=0$ [A1.1]. The height of the contour for the hydrostatic stress is in this case

$$\omega = \frac{(1+\nu)^2}{4\sqrt{3}\pi}\left(\frac{K_\mathrm{I}}{\sigma_h}\right)^2 \tag{A1.3.1}$$

For the computation of the swelling zones ahead of very *short surface cracks*, $\omega \approx a$, a Finite Element analysis was performed. Fig. A1.7 shows such a crack together with contours for constant hydrostatic stresses σ_h, which under "weak swelling" conditions describe the shape of the swelling zones (dashed contours). This treatment is similar to the case of "weak phase transformation" assuming that the stress field caused by a phase transformation, e.g. in zirconia ceramics, does not affect the stress field triggering the transformation [A1.1, A1.6].

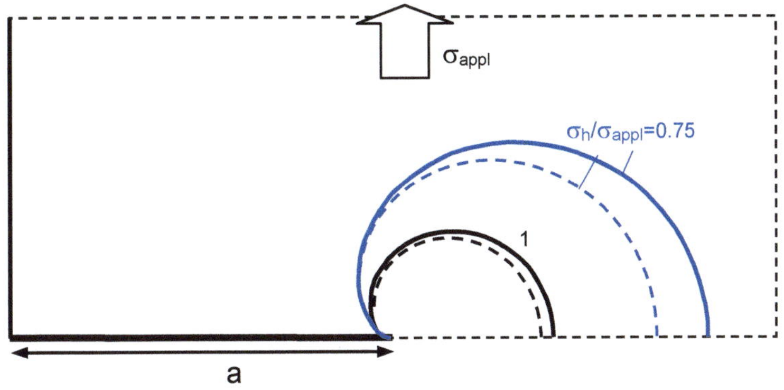

Fig. A1.7 Contours of hydrostatic stress fields ahead of a short crack with depth *a* in a semi-infinite plate loaded by a tensile stress σ_{appl} (solid curves). The dashed contours show the related long-crack solution with $\omega/a\to 0$.

The solid contours in Fig. A1.7 show the contours of hydrostatic stresses σ_h for a tensile stress σ_{appl}. For a crack of depth *a*, the "applied stress intensity factor" results as

$$K = 1.122\sigma_{appl}\sqrt{\pi a} \tag{A1.3.2}$$

In Fig. A1.8 the zone heights from FE are compared with the near-tip stresses as resulting from the long-crack solution

$$\sigma_h = \tfrac{2}{3}(1+\nu)\frac{K}{\sqrt{2\pi r}}\cos(\varphi/2)\,, \quad r = \frac{\omega}{\sin \pi/3} \tag{A1.3.3}$$

For $\omega/a \leq 1/4$, differences are negligible.

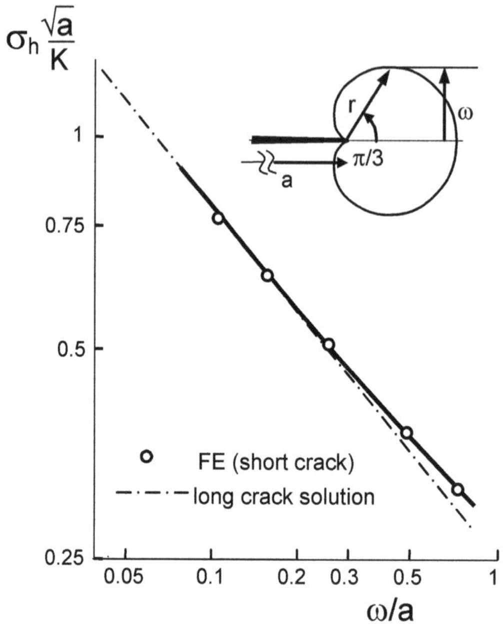

Fig. A1.8 Hydrostatic stress at a crack under tension, comparison of short-crack results obtained via FE with the solution for a long crack according to McMeeking and Evans [A1.1]; contour for constant hydrostatic stress as insert.

A1.4 Diffusion-controlled heart-shaped swelling zones

A1.4.1 Zone shape

The diffusion coefficient, D, which determines the rate of water diffusion into a solid depends on the stresses in the material and can be expressed by

$$D = D_0 \exp\left(\frac{\sigma_h \Delta V_w}{RT}\right) \tag{A1.4.1}$$

where σ_h is the hydrostatic stress, T the absolute temperature, R the universal gas constant, and ΔV_w the activation volume for the diffusion of water in silica. Stress-enhanced diffusion occurs especially in the singular stress field of mechanically loaded cracks where $\sigma \rightarrow \infty$.

If K denotes the stress intensity factor at the tip of a crack, then the singular near-tip hydrostatic stress field is given by the following equation:

$$\sigma_h = \tfrac{2}{3}(1+\nu)\frac{K}{\sqrt{2\pi r}}\cos(\varphi/2) \tag{A1.4.2}$$

where r and φ are the the polar coordinates with origin at the crack tip. This results in heart-shaped diffusion and swelling zones around the crack tip, Fig. A1.9.

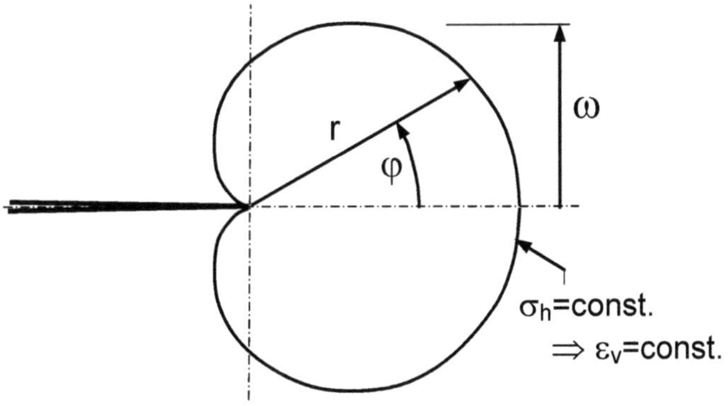

Fig. A1.9 Contour line for constant hydrostatic stress σ_h.

A1.4.2 Shielding stress intensity factor

In order to compute the shielding stress intensity factor caused by an arbitrary radial distribution $\varepsilon_v(r)$, the case of a constant swelling strain may be used, Fig. A1.10a. For this zone, McMeeking and Evans [A1.1] showed that

$$K_{sh} = 0 \tag{A1.4.3}$$

Since this result does not depend on the zone height, ω, it results for variable swelling strains [A1.2]

$$K_{sh} = \int_0^{\omega} \varepsilon_v(\omega')\left(\frac{dK_{sh}(\omega')}{\varepsilon_v\, d\omega'}\right)_{\varepsilon=const} d\omega' = 0 \tag{A1.4.4}$$

i.e. eq.(A1.4.3) holds also in the case of any radial distribution of swelling, Fig. A1.10b.

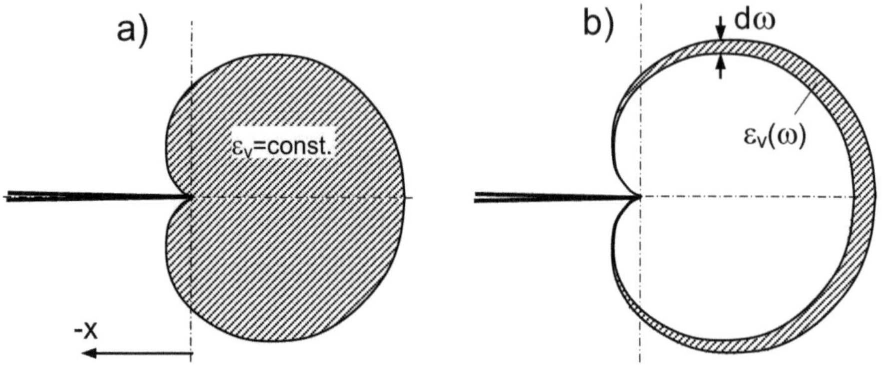

Fig. A1.10 a) Swelling zone with constant strain ε_v, b) zone increment with local strain $\varepsilon_v(\omega)$.

A1.5 Shielding stress intensity factor for cracks in a half-space

A1.5.1 Edge crack

Finite Element computations were carried out for a small edge crack in a rectangular bending bar, a/W=0.005, under plane strain conditions, Fig. A1.11a. The plate height was chosen as H/W=2 representing sufficiently the infinitely long bar. Swelling was assumed taking place from both specimen side-surfaces. Figure A1.11b represents the shielding stress intensity factor as a function of b/a. The FE results are introduced as the circles.

In order to obtain an appropriate fitting relation the analytically known limit values have to be included.

For $b/a \rightarrow 0$ it was derived in [A1.7]:

$$K_{sh} = -\psi \frac{\varepsilon_v E}{1-\nu}\sqrt{b} - \frac{\varepsilon_v E}{3(1-\nu)} \frac{1.465b}{\sqrt{a}} \tag{A1.5.1}$$

with $\psi \cong 0.255$. The first term reflects the swelling zone along the crack surface. Its value was given by McMeeking and Evans [A1.1] and determined in [A1.7] also by FE-computations.

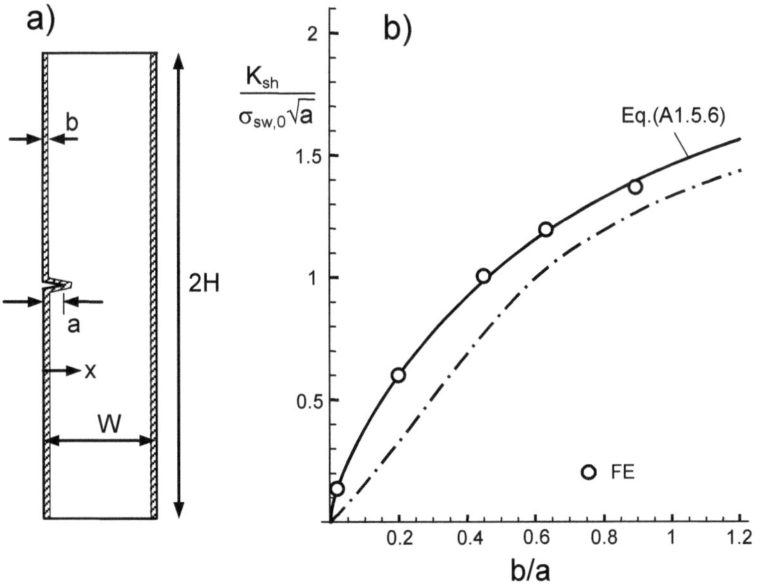

Fig. A1.11 a) Specimen dimensions, b) Finite Element results (circles) compared with the interpolation relation of Eq.(A1.5.6), dash-dotted curve: contribution of the side-surface swelling zone.

Under the assumption of an erfc-shaped swelling strain distribution extending from the specimen surface, the swelling stresses in x-direction are

$$\sigma_n \cong \sigma_{sw,0}\,\mathrm{erfc}\left[\frac{x}{2b}\right] \qquad \text{(A1.5.2)}$$

with the compressive surface stress $\sigma_{sw,0}<0$. The contribution of this layer to the shielding stress intensity factor can be computed as

$$K_{sh,surface} = \int_0^\infty \sigma_n h(x,a)dx \qquad \text{(A1.5.3)}$$

with the normal stress σ_n on the crack and the weight function h as given e.g. in [A1.4]. Figure A1.12 shows h for a crack in the half-space. For very thin swelling zones, $b/a \ll 1$, it holds $h=1.465/\sqrt{a}$.

The shielding stress intensity factor $K_{sh,surface}$ for $b \ll a$ then reads

$$K_{sh,surface} \overset{b/a \to 0}{=} \sigma_{sw,0} \frac{2}{\sqrt{\pi}} \frac{1.465\,b}{\sqrt{a}} \qquad \text{(A1.5.4)}$$

which is about 12% larger than the solution for a step-shaped swelling distribution. For the limit $b/a>>1$ it holds

$$K_{sh} = \sigma_{sw,0} F \sqrt{\pi\, a} \qquad \text{(A1.5.5)}$$

with F=1.1215. The solid curve in Fig. A1.11b represents an interpolation relation including the limit cases that fits best to the FE-results within $0 \le b/a \le 1$:

$$K_{sh} \cong 1.122\sqrt{\pi a}\,\sigma_{sw,0} \tanh^{3/2}\left([0.385\sqrt{\frac{b}{a}} + 0.832\frac{b}{a}]^{2/3} \right) \qquad \text{(A1.5.6)}$$

The dash-dotted curve in Fig. A1.11b represents the contribution caused by the side-surface swelling zone exclusively. For $b/a>1$ this contribution counts for more than 90% of the total shielding stress intensity factor.

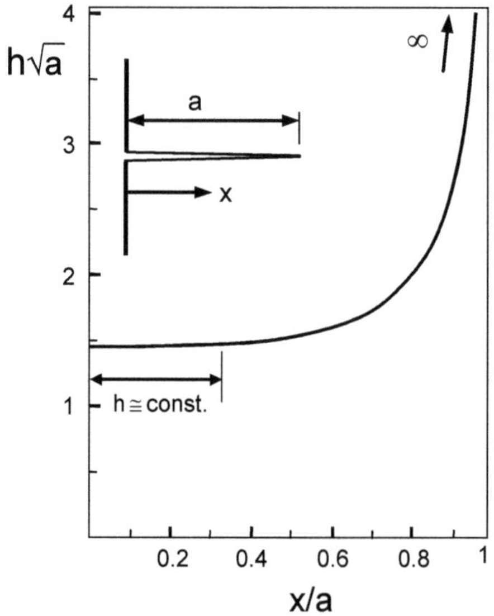

Fig. A1.12 Weight function h for an edge-cracked half-space.

A1.5.2 Semi-circular crack

The geometry and shielding stress intensity factors from FE-computations on a semi-circular crack in the semi-infinite body are shown in Fig. A1.13.

The asymptotic limits are well known (upper limit for point B: 1.29, for point A: 1.17). The bold interpolation curves are given as

$$K_{sh,B} \cong 1.29\sqrt{a}\,\sigma_{sw,0} \tanh\left(1.327\sqrt{\frac{b}{a}} + 0.064\frac{b}{a}\right) \tag{A1.5.7}$$

$$K_{sh,A} \cong 1.17\sqrt{a}\,\sigma_{sw,0} \tanh\left(0.698\sqrt{\frac{b}{a}} + 0.317\frac{b}{a}\right) \tag{A1.5.8}$$

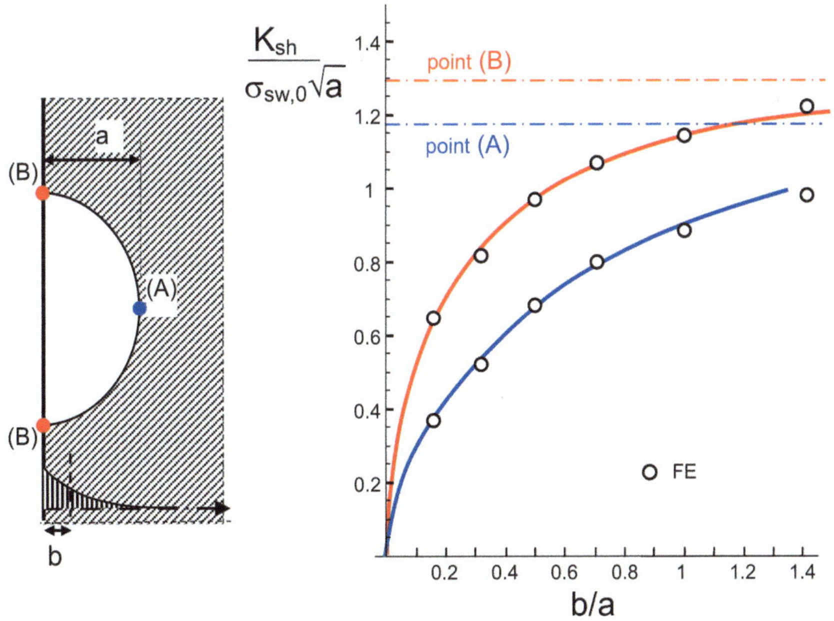

Fig. A1.13 Shielding stress intensity factors at the deepest point (A) and the surface points (B) of a semi-circular crack in a semi-infinite body as a function of the ratio of diffusion length *b* to the crack depth *a*.

A1.5.3 Semi-elliptical crack

Two contributions make up the shielding stress intensity factor: one coming from the diffusion zone originating from the crack faces, $K_{sh,1}$, the second originating from the external surfaces of the specimen, $K_{sh,2}$. The total shielding stress intensity factor is the sum of both of these:

$$K_{sh,A,B} = K_{sh,1} + K_{sh,2} \tag{A1.5.9}$$

Crack-face zone: The diffusion zone developing from the crack faces is given by the following equation [A1.1], which holds both at point *A* and point *B*:

$$K_{sh,1} = -\psi \frac{\varepsilon_v E}{1-\nu}\sqrt{b} \tag{A1.5.10}$$

The shielding stress intensity factors for points (A) and (B) of the semi-elliptic crack, $K_{sh,2}$, are plotted in Fig. A1.14 versus a/c and b/a. It is obvious that the shielding stress intensity factor at the surface points (B) is roughly independent of a/c for about $a/c>0.6$. For small b/a, the stress intensity factor $K_{sh,A,2}$ is much less than $K_{sh,B,2}$. This solution plotted in Fig. A1.14 can be written as

$$K_{sh,A,B,2} = \sigma_0\sqrt{\pi a}\, F_{A,B} \frac{1.13 - 0.09\sqrt{\frac{a}{c}}\left\{\tanh\left[\left(\frac{a}{c}\right)^{10}\right]\right\}^{1/20}}{\sqrt{1+1.464\left(\frac{a}{c}\right)^{1.65}}} \tag{A1.5.11}$$

where

$$F_A = \sum_{n=1}^{3} \frac{B_n}{1 + A_1 n\frac{a}{2b} + A_2\left(n\frac{a}{2b}\right)^2} \tag{A1.5.12}$$

and

$$F_B = \sum_{n=1}^{3} 1.1\sqrt{\frac{a}{c}} \frac{B_n}{1+\left(0.1186 + 0.128\frac{a}{c} - 0.1134\frac{a^2}{c^2} + 0.0288\frac{a^3}{c^3}\right)n\frac{a}{2b}} \tag{A1.5.13}$$

$$A_1 = 0.43 + 0.266\left(\frac{a}{c}\right)^{0.42} \quad \text{(A1.5.14)}$$

$$A_2 = 0.266\left(\frac{a}{c}\right)^{0.42} \quad \text{(A1.5.15)}$$

B_1=−0.37664, B_2=2.7314, B_3=−1.35521

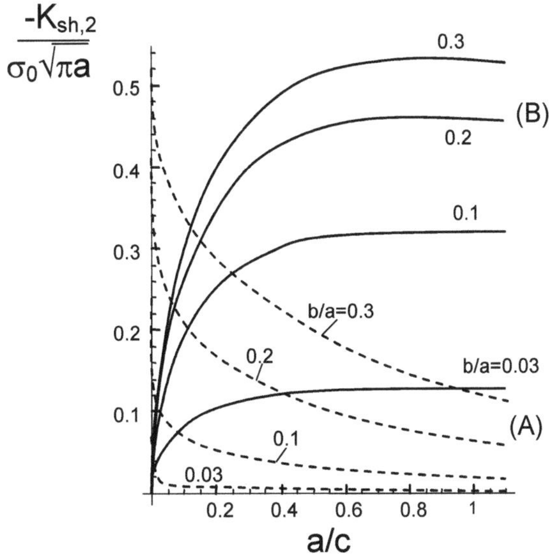

Fig. A1.14 Influence of aspect ratio *a*/*c* and relative depth of the swelling zone *b*/*a* on the shielding stress intensity factors $K_{sh,A}$ at the deepest and $K_{sh,B}$ at the surface points of a semi-elliptical surface crack.

A1.6 Deformed steady-state swelling zones

The computations of the swelling zones are mostly carried out for diffusion in a motionless coordinate system with a fixed crack tip, i.e. for an *arrested* crack. Whereas for a *growing* crack the transversal diffusion normal to the crack plane is hardly affected by the moving crack tip, the diffusion and the crack propagation compete in crack direction [A1.8, A1.9]. The consequence for strongly stress enhanced diffusion zones with cordial shape is a reduced ratio of the zone size in length direction r_0 to the zone height ω (Fig. A1.15), which for an arrested crack is [A1.10]

$$\frac{r_0(0)}{\omega(0)} = \frac{8}{\sqrt{27}} \tag{A1.6.1}$$

($r_0(0)$, $\omega(0)$ stands for $r_0(v=0)$ and $\omega(v=0)$, respectively). For the stress intensity factor of the deformed zones one can use some single results from literature.

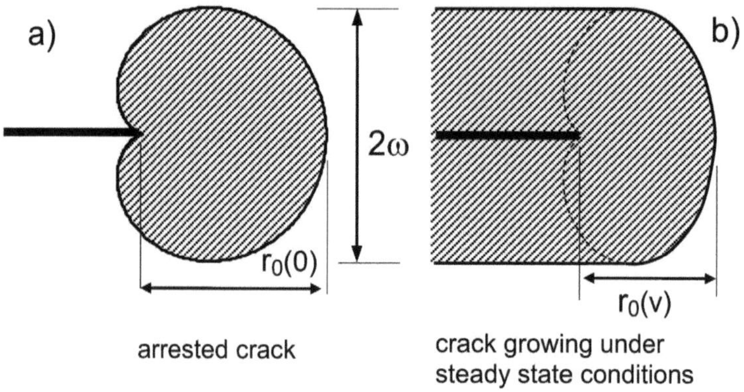

Fig. A1.15 a) Swelling zone ahead the tip of an arrested crack, b) expected reduced zone length for a crack growing at a constant crack-growth rate v (schematic).

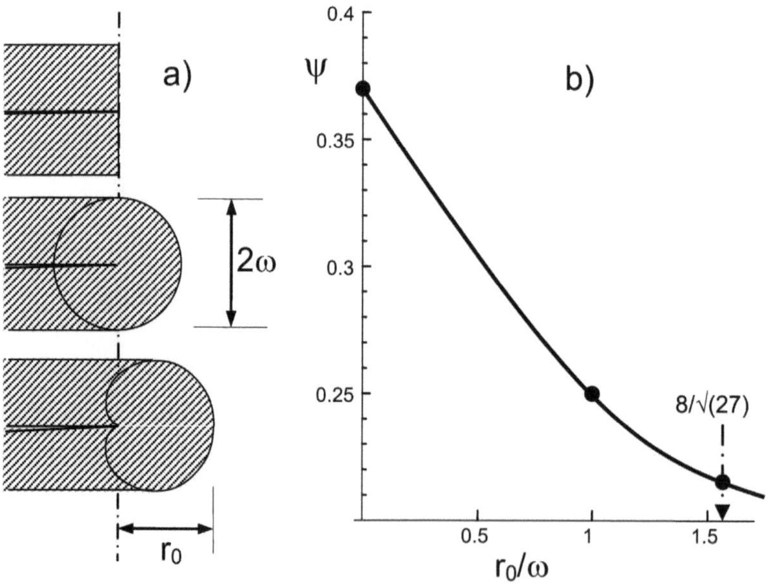

Fig. A1.16 a) Swelling zones with different shape at the zone end, b) shielding stress intensity factors by eq.(A1.7.1) for the zones given in a).

In [A1.11] it was found for crack rates $v>10^{-12}$ m/s under the assumption that the diffusion normal to the crack-plane direction is not affected by the crack rate ($v_\perp=0$):

$$\tfrac{3}{4}\frac{8}{\sqrt{27}} < \frac{r_0(v)}{\omega(v)} < \frac{8}{\sqrt{27}} \tag{A1.6.2}$$

Finally, Fig. A1.16 shows the coefficient ψ for shielding stress intensity factors, eq.(A1.5.10).

A1.7 Edge-crack passing and leaving a swelling zone

During the soaking treatment under tension the crack-tip region is surrounded by a swelling zone. The shielding stress intensity factor K_{sh} caused by this zone was given in [A1.1] according to eq.(A1.5.10)

$$K_{sh} = -\psi \frac{\varepsilon_v E}{1-\nu}\sqrt{\omega} \tag{A1.7.1}$$

with the coefficient ψ depending on the zone shape in the crack-tip region.
When crack propagation has taken place during soaking, the swelling zone extends with crack length as illustrated in Fig. A1.17b. For the case of weak swelling as discussed by McMeeking and Evans [A1.1] the heart-shaped contour remains ahead of the tip and the coefficient is $\psi=0.22$. After high-temperature soaking, these swelling zones are maintained at room temperature where the strength tests are performed.

When the crack undergoes continuously increasing loading, the crack must extend and will escape of the initial swelling zone. Concequently, the coeffient ψ in eq.(A1.7.1) must become dependent on the crack extension Δa in the strength test. The red curve in Fig. A1.17c represents analytical results for a heart-shaped swelling zone we got by application of the McMeeking and Evans [A1.1] procedure. A maximum of ψ is reached at $\Delta a/\omega = 8/\sqrt{27} \cong 1.54$ with $\psi\approx 0.3$. In the case of a grown crack with extended swelling zone (black curve) it results $\psi\approx 0.33$.

At higher temperatures and after long time, also unloaded cracks may develop a swelling zone of thickness b by water diffusion into the crack faces. Figure A1.18a represents a crack with a circular zone front in the crack tip region.

If such a crack grows by an amount of Δa, it will escape from the initial swelling zone, as illustrated in Fig. A1.18a-Fig. A1.d for different zone ends. The shielding stress intensity factors for two limit cases of the zone shape near the crack tip are plotted in Fig. A1.18e. For the circular zone end (Fig. A1.18b), the shielding stress intensity factor is given by the squares [A1.12]. A straight zone end with r_0/b=0 yields the dash-dotted curve. The two dependencies can be described by eq.(A1.7.1) where now ψ is a function of $\Delta a/b$.

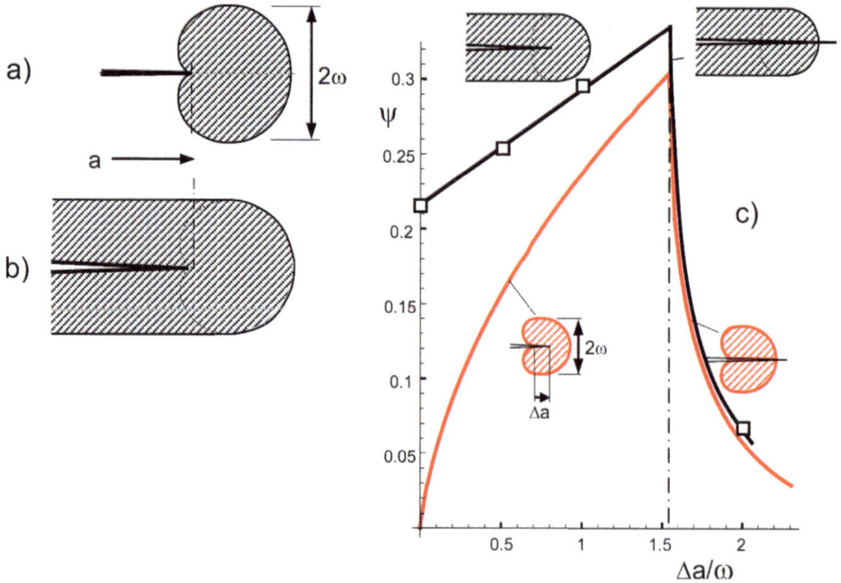

Fig. A1.17 a) Swelling zone in the absence of crack propagation, b) grown crack for weak swelling, c) shielding stress intensity factor for crack extension through the initially swelling zone, red curve: crack passing the initial zone developing for an arrested crack, obtained by application of the procedure by McMeeking and Evans [A1.1]; black: grown crack after additional extension Δa obtained from FE-analysis, [A1.12].

The stress intensity factor solution for the case r_0/b=½ (Fig. A1.18d) can be obtained by linear interpolation of the two limit cases resulting in the dashed curve of Fig. A1.18e. The coefficient ψ may be approximated by

$$\psi \cong \begin{cases} 0.305 + \frac{1}{12}\Delta a/b & for \quad \Delta a/b \le \frac{1}{2} \\ 0.34\exp[-2\sqrt{\Delta a/b - \frac{1}{2}}\ ] & \quad \Delta a/b > \frac{1}{2} \end{cases} \tag{A1.7.2}$$

It should be mentioned that the curves in Fig. A1.17c and A1.18e change abruptly caused by the assumption of a sharp layer boundary. For an arbitrarily varying strain distribution, the weight function or Green's function method from [A1.2] can be used.

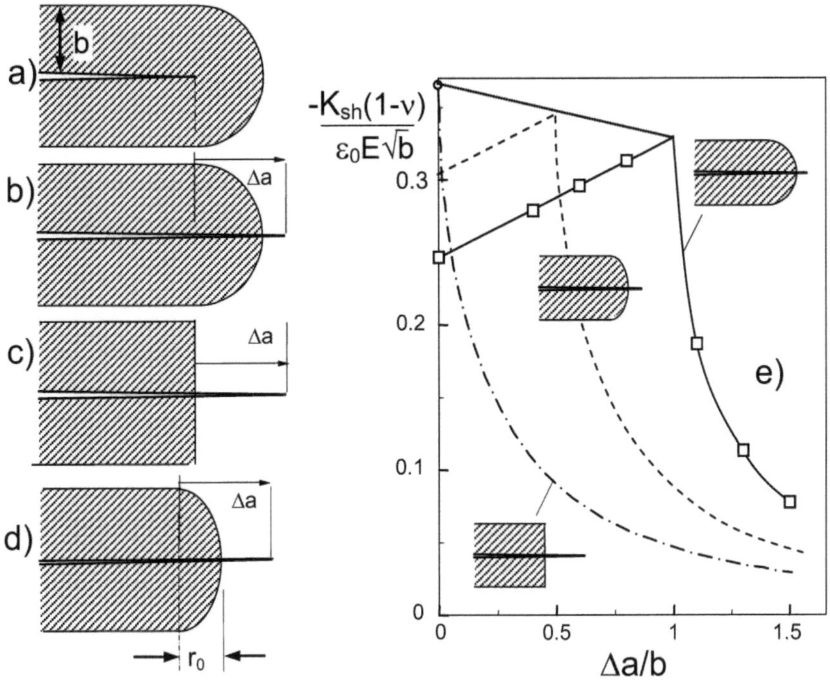

Fig. A1.18 a) Crack with a thick circular swelling layer of constant swelling strain, b) crack after additional extension Δa for a circular zone, c) for a rectangular zone, d) for a zone with $r_0/b=1/2$, e) variation of the shielding stress intensity factor for a crack growing in and outside the initial layer according to [A1.10] (dash-dotted curve) and [A1.12] (squares); dashed curve: interpolated from the continuous and dash-dotted curves for $r_0/b=1/2$.

The stress intensity factor caused by a swelling layer of thickness db' in the surface distance $z=b'$ (Fig. A1.19a) undergoing the strain $\varepsilon(b')$ results as

$$d(K_{sh}) = K_{sh}(b'+db') - K_{sh}(b') \qquad \text{(A1.7.3)}$$

This gives rise for a Green's- or "weight function" h defined as

$$h = \frac{d(K_{sh})}{db'} \tag{A1.7.4}$$

For any swelling distribution ε(*z*)=ε(*b*'), the related shielding stress intensity factor is then obtained by summing over all the zone increments d*b*' multiplied with the local swelling strain

$$K_{sh} = \frac{1}{\varepsilon_0}\int_0^\infty \varepsilon(b')\,h(b')\,db', \quad \varepsilon_0 = \varepsilon(b'=0) \tag{A1.7.5}$$

In the absence of stress-enhanced diffusion, the swelling profiles are represented by the complementary error function. The volumetric strain at the location *z*=*b*' is in this case

$$\varepsilon = \varepsilon_0 \operatorname{erfc}\left(\frac{b'}{2\sqrt{D_0 t}}\right) \tag{A1.7.6}$$

where ε_0 is the strain at the free surface, D_0 the diffusion coefficient and *t* the time after the first water contact.

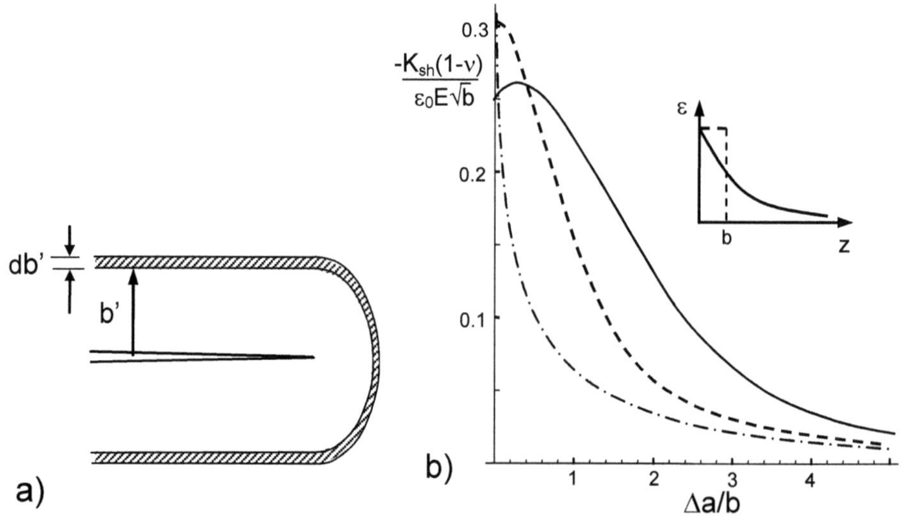

Fig. A1.19 a) Green's function as the shielding stress intensity factor of a thin swelling layer of thickness *b*', b) stress intensity factors according to the dashed curve in (d) for an erfc-shaped strain distribution (insert); curves according to Fig. A1.18e.

Equation (A1.7.5) has been evaluated for the case of $r_0/b=½$. The resulting shielding stress intensity factors related to the erfc-shaped swelling profiles are shown in Fig. A1.19b. The numerical result for $r_0/b=½$ can be approximated by equation (A1.7.1) with

$$\psi \approx \frac{0.305}{1+\left(\dfrac{\Delta a}{b}\right)^2} \qquad \text{(A1.7.7)}$$

References

A1.1 R.M. McMeeking, and A.G. Evans, "Mechanics of Transformation-Toughening in Brittle Materials," *J. Am. Ceram. Soc.* **65** 242-246 (1982).

A1.2 Wiederhorn, S.M., Fett, T., Rizzi, G., Hoffmann, M.J., Guin, J.-P., The Effect of Water Penetration on Crack Growth in Silica Glass, Engng. Fract. Mech. **100** (2013), 3-16.

A1.3 O.L. Bowie, Solutions of plane crack problems by mapping technique, in: Mechanics of Fracture (Ed. S.C. Sih), Vol. 1, (1973), 1-5, Noordhoff International Publishing, Leyden.

A1.4 Fett, T., Munz, D., Stress intensity factors and weight functions, Computational Mechanics Publications, Southampton, 1997.

A1.5 M. Isida and H. Tsuru, *Transactions Japan Society of Mechanical Engineers* **49** (1983) 641-645.

A1.6 Budiansky, B., Hutchinson, J.W., Lambropoulos, J.C., Continuum theory of dilatant transformation toughening in ceramics, Int. J. Solids and Structures, **25**(1989), 635-646.

A1.7 S.M. Wiederhorn, T. Fett, G. Rizzi, M. Hoffmann, J.-P. Guin, "Water Penetration – its Effect on the Strength and Toughness of Silica Glass," *Met. Mater. Trans. A*, **44** (2013) [3], 1164 -1174.

A1.8 Zhao, T., Nehorai, A., Detection and localization of a moving biochemical source in a semi-infinite medium, Report, Electrical and Computer Engineering Dept. Univ. of Illinois, Chicago, USA.

A1.9 Babich, V.M., On asymptotic solutions of the problems of moving sources of diffusion and oscillations, UDC 534.231.1, 517.226, 1991, pp. 3067-3071, Plenum Publishing Corp.

A1.10 McMeeking, R.M., Evans, A.G., Mechanics of Transformation-Toughening in Brittle Materials, J. Am. Ceram. Soc. **65** (1982), 242-246.

A1.11 T. Fett, New contributions to R-curves and bridging stresses – Applications of weight functions, IAM 3, KIT Scientific Publishing, 2012, Karlsruhe.

A1.12 Fett, T., Guin, J.P., Wiederhorn, S.M., Interpretation of effects at the static fatigue limit of soda-lime-silicate glass, Engng. Fract. Mech. **72** (2005), 2774-279.

A2 Stresses at pores

In order to allow application of continuum mechanics relations in terms of stresses and strains, the complicated SiO_2 micro structure is simplified as shown in Fig. A2.1. Figure A2.1a shows a volume element of the silica network near the free surface containing SiO_2 rings of different size. The largest ones are indicated by the dashed circles. Figure A2.1b illustrates the simplification by spherical pores.

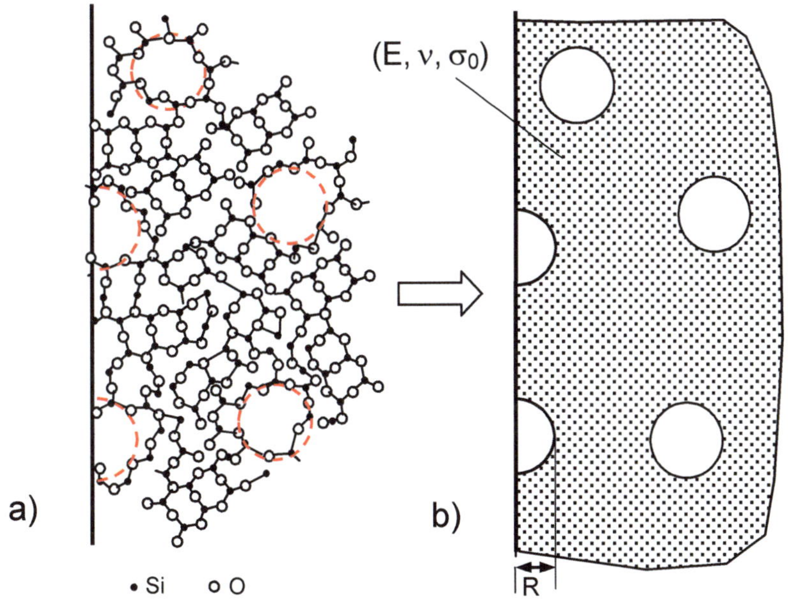

Fig. A2.1 a) Silica network, b) large SiO_2 rings replaced by nano-pores.

A2.1 Two-dimensional array of half-spherical surface pores

Fig. A2.2a shows a symmetry element of a two-dimensional array of semi-spherical pores at a flat surface. The radius of the half-sphere is R and the distance to the next one $2H$.

The applied stress is denoted as σ_∞. The stresses σ_y along the equator between points (A) and (B) are represented in Fig. A2.2b for several values of *H*/*R* and Poisson's ratio ν=0.17. Fig. A2.2c shows the reciprocal stress at the surface and the deepest point as a function of *R*/*H*.

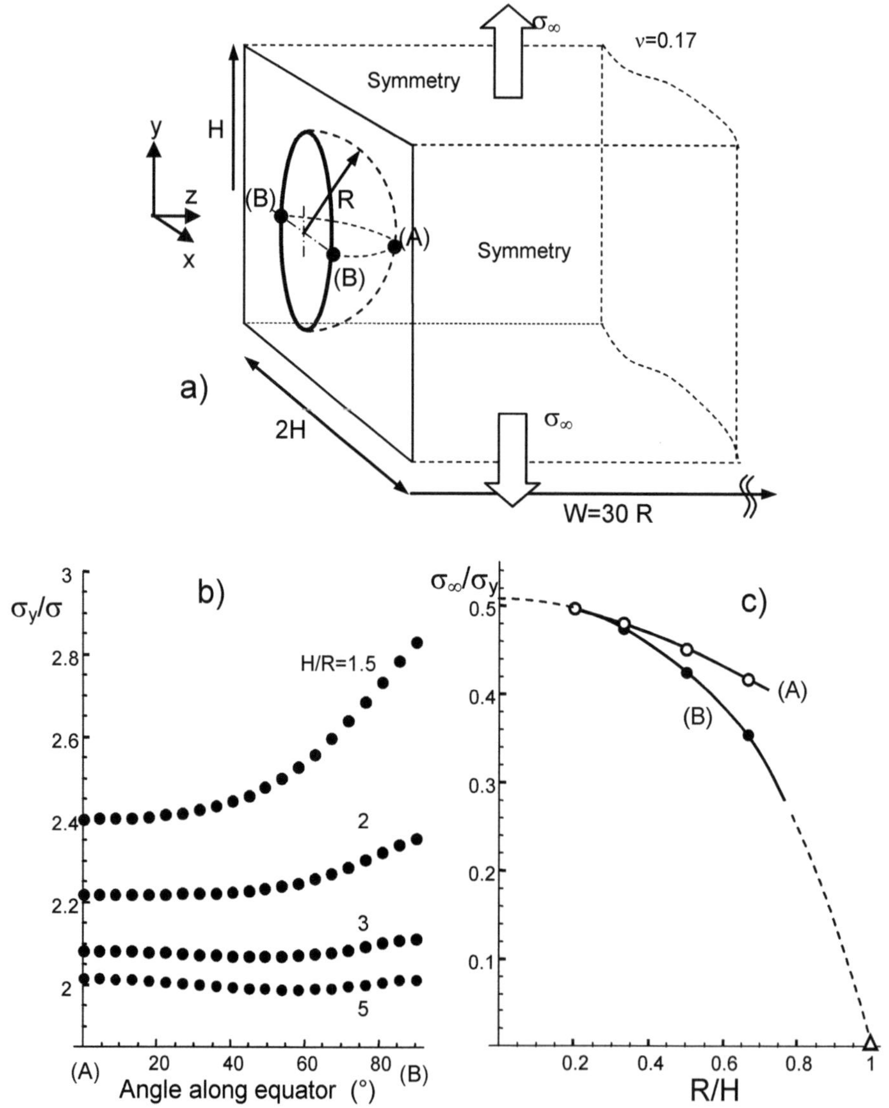

Fig. A2.2 a) Semi-spherical surface pore, b) stress distribution σ_y along the equator, c) reciprocal stresses, $1/\sigma_y$ at the surface (B) and the deepest point (A), ν=0.17.

The stresses σ_y at the surface points, (B), and the deepest point (A) can be approximated for $R/H<0.8$ by

$$\left.\frac{\sigma_y}{\sigma_\infty}\right|_{(B)} = \frac{1}{0.51 - 0.254\left(\frac{R}{H}\right)^2 - 0.255\left(\frac{R}{H}\right)^4} \tag{A2.1.1}$$

$$\left.\frac{\sigma_y}{\sigma_\infty}\right|_{(A)} = \frac{1}{0.503 - 0.197\left(\frac{R}{H}\right)^2} \tag{A2.1.2}$$

A2.2 Broken bonds as tablet-shaped nano-pores

In the preceding considerations, it was implicitly assumed that no interaction of the bond-breaking events is present and consequently the hydroxyl concentration was considered to be constant over the cross section. For very high stresses, this cannot be guaranteed. On one hand side, a stress concentration occurs around a single broken bond since the surrounding bonds must transfer the load that was originally carried by this bond, Fig. A2.3. The increased load in the neighboured bonds causes an increased failure probability by stress-enhanced hydroxyl generation. The result would be an uneven distribution of the hydroxide content on the molecular level.

A SiO_2 ring of the silica network may be described by a spherical (nano)-void in a continuous material (left part of Fig. A2.3). This allows a fracture mechanics treatment by an FE-study. According to the finite crack-tip curvature for cracks, the radius of such a void has to be expected in the order of $\rho = 0.5$ nm [A2.1].

Under a load in *z*-direction, the red Si-O-bond may be cracked by reaction with molecular water (blue/red). Since the open bond cannot transfer load, the void now acts as a disk-like defect of length $2(\rho+L)$ oriented perpendicular to the load direction. The stress increase $\sigma_{max}/\sigma_\infty$ at the notch root is shown in Fig. A2.4a as a function of the length L (circles). A rough representation of these data may be given by the simple relation

$$\frac{\sigma_z}{\sigma_\infty} = 1.48 \times \sqrt{1.84 + \frac{L}{\rho}} \tag{A2.2.1}$$

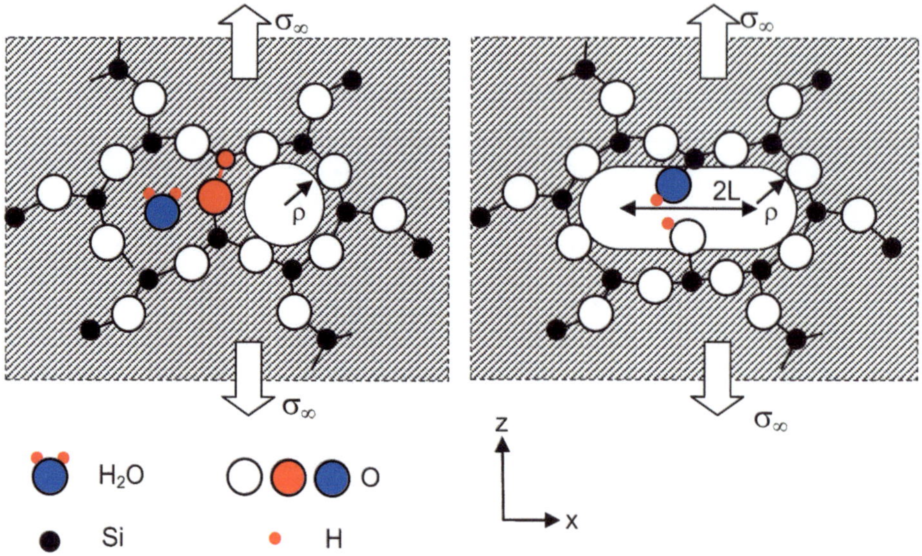

Fig. A2.3 Left: Bond breaking under uni-axial tension by reaction with molecular water (attacked SiO-bond in red); right: open bond represented by a disk-like notch of diameter $2(\rho+L)$.

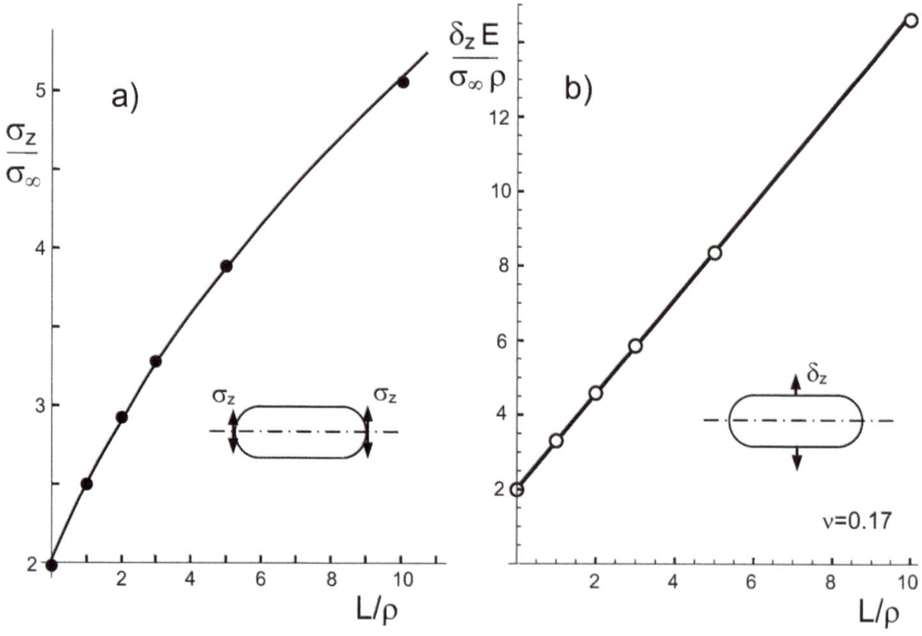

Fig. A2.4 a) Stress concentration at the notch root for several values of notch length L (circles: FE-results, curve: approximation by eq.(A2.2.1), b) displacements in the axis direction.

The maximum displacement in the centre is given in Fig. A2.4b. The straight-line dependency is represented by

$$\delta_z \frac{E}{\rho\sigma_\infty} = 2 + 1.275\frac{L}{\rho} \tag{A2.2.2}$$

Reference

A2.1 S.M. Wiederhorn, E.R. Fuller, Jr. and R. Thomson, "Micromechanisms of crack growth in ceramics and glasses in corrosive environments," Metal Science, **14**(1980), 450-8.

A3 Effect of surface cracks on the diffusion profile

A3.1 Crack modelling

In ground silica surfaces cracks are present even in the case of finishing by mechanical polishing or chemical etching. This has been outlined by Suratwala et al. [A3.1]. Such a surface state is schematically shown in Fig. A3.1a in side- and front view. The individual distributions of the depths a and the widths L are approximated in [A3.1] by exponential distribution functions.

For a simple explanation of the effect of such cracks on diffusion, let us replace the distributions of a and L by periodically arranged cracks, Fig. A3.1b, of average crack surface

$$S = 2\langle a \times L \rangle \tag{A3.1.1}$$

in the surface element of length A and width B, where the factor 2 counts for two crack faces of each crack. The angle brackets $\langle \ldots \rangle$ indicate averaging.

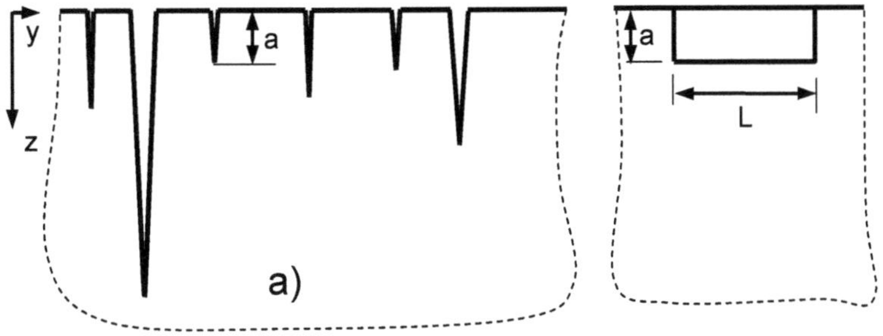

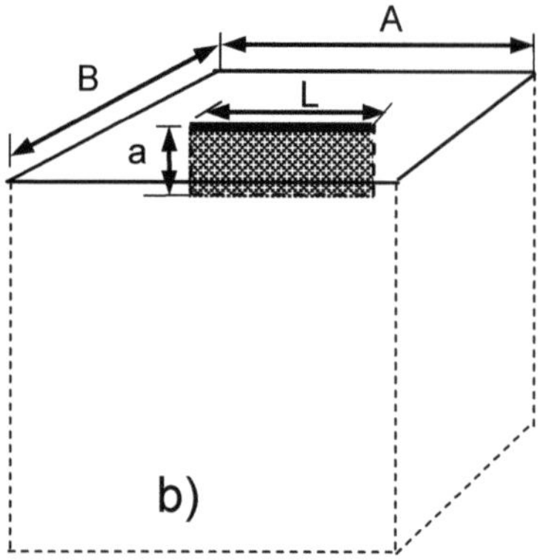

Fig. A3.1 a) Surface cracks due to surface treatment, front and side view, b) cracks of a) replaced by a periodical element with surface area $A\times B$ containing a single crack of mean crack surface S.

A3.2 Diffusion profiles

When the "unit cell" in Fig. A3.1b is exposed to a humid environment, water can diffuse into the whole surface given by the original surface area $A\times B$ and the crack surface S, $A\times B+S$, shown in Fig. A3.3a. The concentration of the water below the original surface, $C(z)$, is given by the complementary error function

$$C(z) = C_0 \,\mathrm{erfc}\left[\frac{z}{2b}\right] \tag{A3.2.1}$$

with the concentration C_0 at the surface. The depth b, the diffusion length, at which the concentration decreased to 50%, depends on diffusivity D and time t via

$$b = \sqrt{Dt} \tag{A3.2.2}$$

The water profile normal to the crack faces is the same with y instead of z in eq.(A3.2.1). Deviations occur in the crack-tip region. In a FE-study [A3.2], the diffusion problem for water-soaked specimens was solved.

The water concentration contours are shown in Fig. A3.2a. It can be seen that the zone size ahead of the tip is reduced. In contrast to the profiles by eq.(A3.2.1), there is also a height reduction visible near the tip.

Figure A3.2b gives the concentration profiles for the *z*-direction ahead of the crack tip and the profile by eq.(A3.2.1) for larger distances behind the tip region. For reasons of simplicity we approximate the water distribution in the diffusion zone by a layer of thickness *b* and a constant water concentration C_0 as indicated by Fig. A3.2b. Near the crack tip, the missing water in *y*-direction is balanced by the water ahead the tip, resulting for C/C_0=0.5 in the red step-shaped distribution in Fig. A3.2a. Figure A3.3a illustrates the water distribution in the unit cell after a short time with $b << a$.

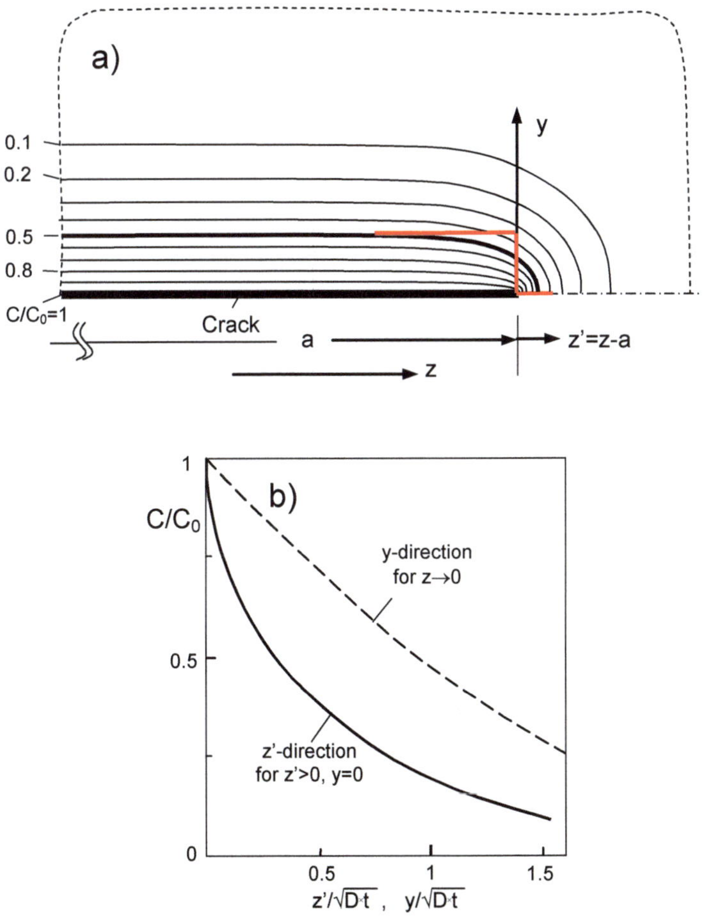

Fig. A3.2 Water diffusion through crack surfaces of a semi-infinite crack, a) contour lines for constant water concentration, b) concentration profiles in length and height directions.

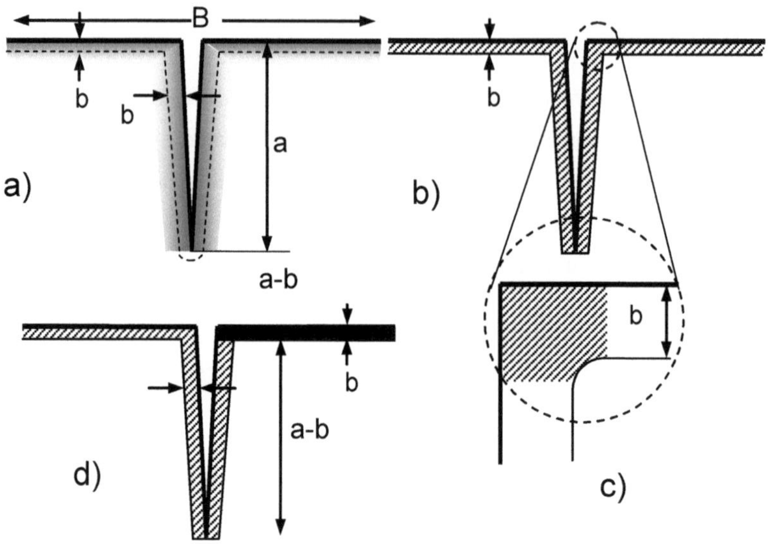

Fig. A3.3 Water diffusion into the crack surface containing a crack of depth *a*; a) water distribution for short times with diffusion zone $b \ll a$, b) water diffusion zone according to eq.(A3.2.1) replaced by a zone of thickness *b* and constant concentration showing the same amount of water, c) interacting diffusion profiles near the crack mouth (schematic), d) separation of overlapping diffusion profiles.

The amount of water, penetrated into the glass is then approximated by a constant concentration of $C=C_0$ in a layer of thickness *b* and $C=0$ outside, Fig. A3.3b. For very short times, the profiles from the surface and from the crack faces interact only at the crack mouth, Fig. A3.3c. As long as this region is small, i.e. for $b \ll a$, it holds for the total amount of water in the unit cell

$$m_C \cong C_0 b[AB + 2La] \tag{A3.2.3}$$

In terms of an apparent layer thickness b_{app} for very short time $t \rightarrow 0$, eq.(A3.2.3) reads

$$b_{app,0} = b_{app,t\rightarrow 0} = b\left(1 + \frac{<S>}{AB}\right) \tag{A3.2.4}$$

For not too large layer thicknesses, let say $b < 1/3$ Min(a, $B/2$), the moderate diffusion interactions may be separated in diffusion layers from the specimen surface over the whole width *B* and from the crack surface over the reduced distance of a-b as illustrated in Fig. A3.3d. In this case we obtain

$$b_{app} \cong b\left(1+\frac{2L}{AB}(a-b)\right)=b\left(1+\frac{<S>}{AB}(1-b/a)\right) \quad \text{(A3.2.5)}$$

This dependency is shown in Fig. A3.4.

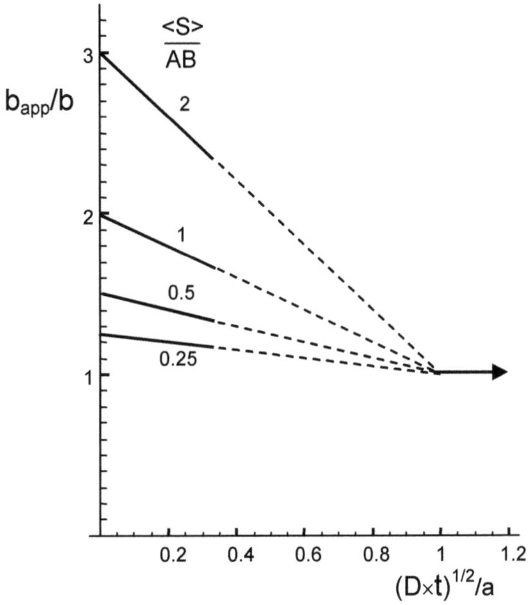

Fig. A3.4 Apparent diffusion layer thickness b_{app} as a function of time.

A3.3 Apparent diffusivity

Equations (A3.2.4) and (A3.2.5) allow the determination of an apparent diffusivity D_{app}, defined via

$$b_{app} \overset{def}{=} \sqrt{D_{app}\,t} \quad \text{(A3.3.1)}$$

This diffusivity is for $b < 1/3 \times \mathrm{Min}(a, B/2)$

$$D_{app} \cong D\left(1+\frac{<S>}{AB}\left(1-\frac{\sqrt{Dt}}{a}\right)\right)^2 \quad \text{(A3.3.2)}$$

Two limit cases can be identified. For very short times, it results from eq.(A3.3.2)

$$D_{app,0} = D_{app,t\to 0} = D\left(1+\frac{<S>}{AB}\right)^2 \tag{A3.3.3}$$

For long times *t* and thick zones *b*, the limit diffusivity is trivially

$$D_{app,\infty} = D_{app,t\to\infty} = D \tag{A3.3.4}$$

Equation (A3.3.2) makes clear that the apparent diffusivity resulting from measurements depends on time as is plotted in the representation of Fig. A3.5.

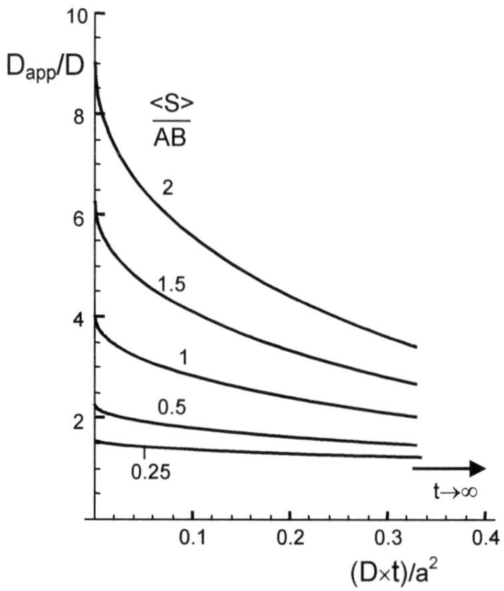

Fig. A3.5 Apparent diffusivity D_{app} as a function of time for several ratios of crack surface <*S*> to specimen surface $A\times B$, arrow indicates the asymptotically reached diffusivity $D_{app\to\infty}$.

A3.4 Interpolation of limit case solutions

Having in mind that the crack sizes and distances are continuously distributed [A3.1], the results from Section A3.2 may be "smoothed" and expressed by interpolation relations. Usual descriptions by exponential functions result in

$$b_{app} \cong b\left(1+\frac{<S>}{AB}\exp(-b/a)\right) \tag{A3.4.1}$$

or

$$b_{app} \cong b\left(1+\frac{<S>}{AB}(1-\tanh(b/a))\right) \tag{A3.4.2}$$

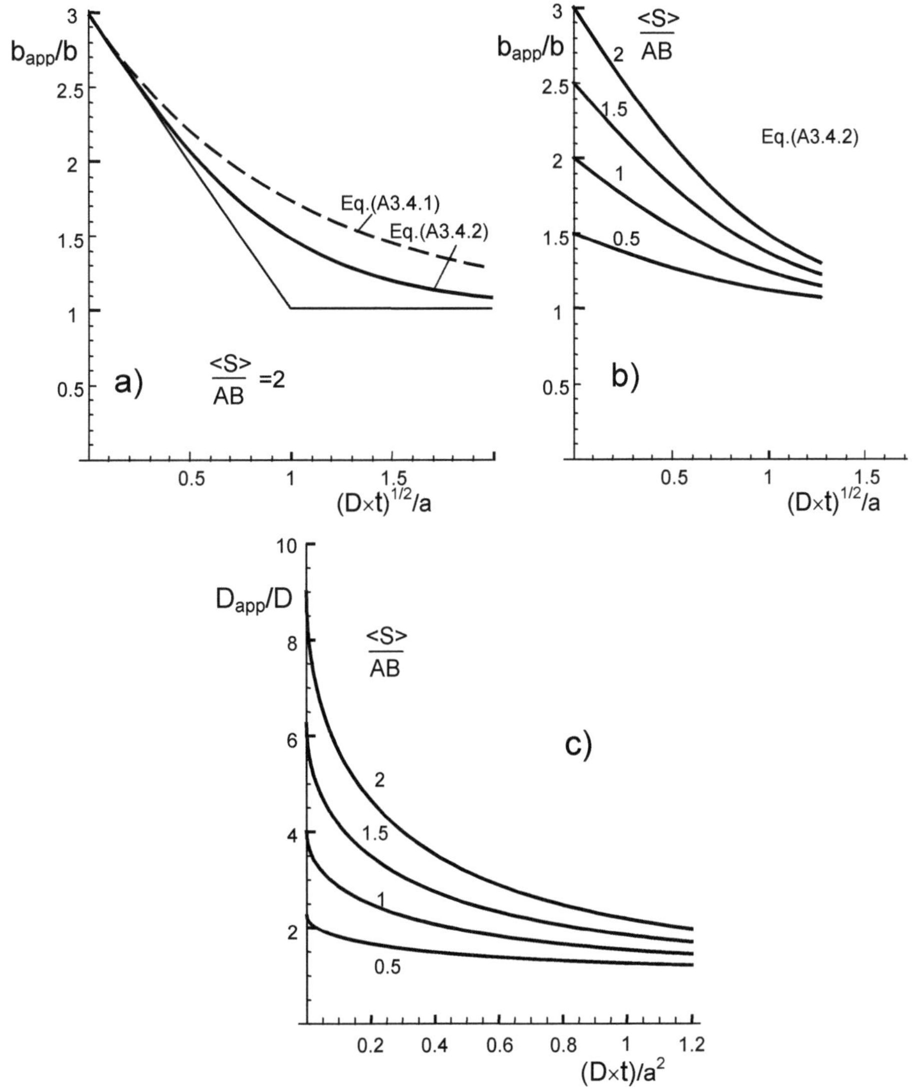

Fig. A3.6 a) Interpolation of limit solutions according to eqs.(A3.4.1) and (A3.4.2), b) interpolations for different crack parameters, c) related apparent diffusivities.

Figure A3.6a shows these interpolations for the case of $<S>/(AB)=2$. The straight lines are again the limit solutions for $t\to 0$ and $t\to\infty$. The dashed curve represents eq.(A3.4.1) and the solid curve eq.(A3.4.2). Use of (A3.4.2) may be recommended. Figure A3.6b shows the interpolations for different parameters $<S>/(AB)$. Finally, Fig. A3.6c gives the apparent diffusivity from eq.(A3.4.2) as

$$D_{app} \cong D\left(1+\frac{<S>}{AB}(1-\tanh(\sqrt{Dt}/a))\right)^2 \tag{A3.4.3}$$

A3.5 Comparison with experimental results

Results of the diffusivity as a function of time are available in the literature. Profiles of time-dependent residual water content by stepwise surface removal were observed by Davis and Tomozawa [A3.3] under water vapour pressure of 355mm Hg at 350°C. Thc results given in Fig. A3.7a, showed reduced normalized diffusion depths with increasing soaking time. This behaviour is plotted again in Fig. A3.7b in terms of an effective diffusivity D_{eff} (red circles).

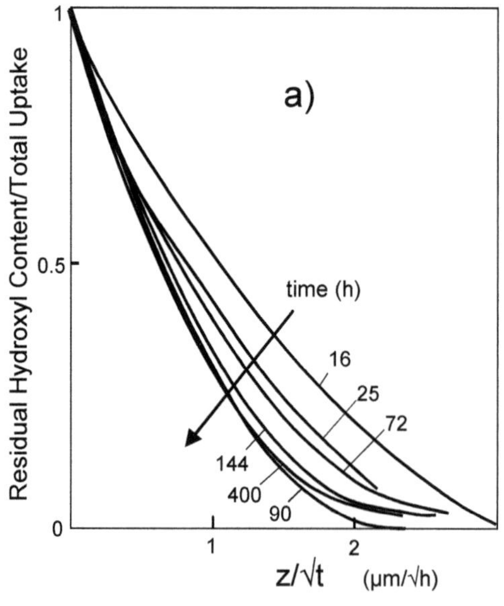

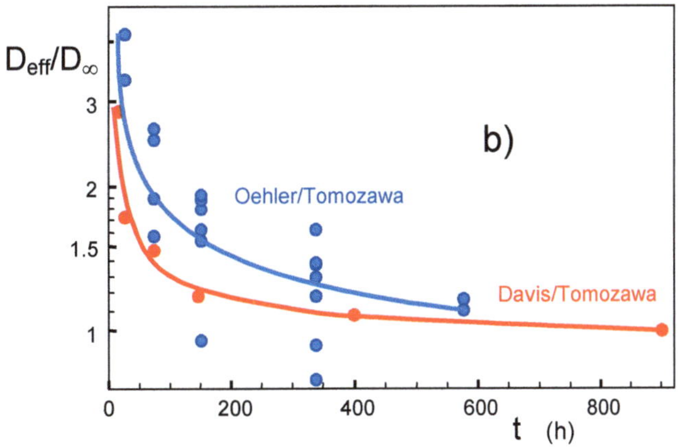

Fig. A3.7 a) Profiles of residual water content during stepwise surface removal by Davis and Tomozawa [A3.3], b) effective diffusivities by Davis and Tomozawa [A3.3] (red circles) and by Oehler and Tomozawa [A3.4] at 250°C and 39 bar water vapour pressure (blue circles).

Measurements of effective diffusivities by Oehler and Tomozawa [A3.4] at 250°C and 39 bar water vapour pressure are introduced by the blue circles. In both cases a clear decrease with time is evident. This effect was discussed in [A3.5] in terms of swelling stresses. In case of the results by Davis and Tomozawa [A3.3] it has to be considered that under the rather low vapour pressure swelling stresses are not very high. As Fig. A3.6c shows, the curves in Fig. A3.7a may also be interpreted as the consequence of surface cracks. That will partially hold for the results in [A3.4], too.

References

A3.1 T. Suratwala, L. Wong P. Miller, M.D. Feit, J. Menapace, R. Steele, P. Davis, D. Walmer, Sub-surface mechanical damage distributions during grinding of fused silica, J. Non-Cryst. Solids, **352** (2006), 5601-5617.

A3.2 S.M. Wiederhorn, T. Fett, G. Rizzi, M. Hoffmann, J.-P. Guin, "Water Penetration – its Effect on the Strength and Toughness of Silica Glass," *Met. Mater. Trans. A*, **44** (2013) [3], 1164 -1174.

A3.3 Davis, K.M., Tomozawa, M., Water diffusion into silica glass: structural changes in silica glass and their effect on water solubility and diffusivity, J. Non-Cryst. Sol. **185** (1995), 203-220.

A3.4 Oehler, A., Tomozawa, M., Water diffusion into silica glass at a low temperature under high water vapor pressure, J. Non-Cryst. Sol. **347** (2004) 211-219.

A3.5 S. M. Wiederhorn, G. Rizzi, G. Schell, S. Wagner, M. J. Hoffmann, T. Fett, Diffusion of water in silica: Influence of moderate stresses, J. Am. Ceram. Soc. **101** (2015), 1180-1190.

A4 Stress reduction and subcritical crack growth

A4.1 Analogy to a mechanical model for subcritical crack growth

The possibility of the breaking of SiO_2 bonds by water molecules without a chemical reaction according to eq.(1.1.1) appears to be quite trivial. The H_2O molecule is retained and only reacts analytically. This was suggested very early on by the authors in a purely mechanical model analysis [A4.1, A4.2, A4.3]. A single bond in front of the crack tip was considered, which had to be broken. The effect of the neighbouring bonds was described as a continuum. The consideration in [A4.1-A4.3] made it possible to explain the subcritical crack growth including the threshold, which, however, also arises freely for the separation process via the chemical reaction [A4.4, A4.5]. The main statements of this analysis will be briefly addressed here.

Figure A4.1a illustrates the isolated bond which is prestressed by the springs (1), reflecting the linear-elastic continuum, Fig. A4.1b. The spring affected by a water molecule is shown in Fig. A4.1c. Figure A4.1d represents two load-displacement curves for bond interactions without (dashed curve) and with water (continuous curve). They are replaced by a non-linear spring (2), i.e. by a non-linear relation between the force F and the bond displacement δ. For the following considerations a Morse relation is used as is common for intermolecular interactions. If F_0 denotes the maximum force, then

$$F = 4F_0[\exp(-s\delta) - \exp(-2s\delta)], \qquad \text{(A4.1.1)}$$

where s is a measure of the range of bond forces. The load–displacement characteristic of a single bond under a thermal activation event is shown by the F-δ diagram in Fig. A4.1d and in Fig. A4.2 for different external loading (increasing load from a) to d)). In the case of a thermal event splitting the bond, the external load becomes reduced with increasing distance of the two bond partners, because the situation shown in Fig. A4.1 is displacement controlled.

Figure A4.2d describes a critical load situation. The intersections $\delta^{(1)}$ and $\delta^{(2)}$ are identical and the straight line is the tangent to the bond force curve. This situation corresponds to catastrophic failure of the bond whereas the tangent in Fig. A4.2b represents a threshold in bond breaking. In Fig. A4.2e the characteristic displacements are related to the macroscopic values of the threshold and the critical stress intensity factor values $K_{\mathrm{I},th}$ and K_{Ic}, and also for the stresses σ_{th} and σ_{c}, respectively. This suggests

$$\frac{\delta}{\delta_c} = \frac{K_{\mathrm{I},th}}{K_{\mathrm{Ic}}} \Rightarrow = \frac{F_{th}}{F_c} = \frac{\sigma_{th}}{\sigma_c}. \tag{A4.1.2}$$

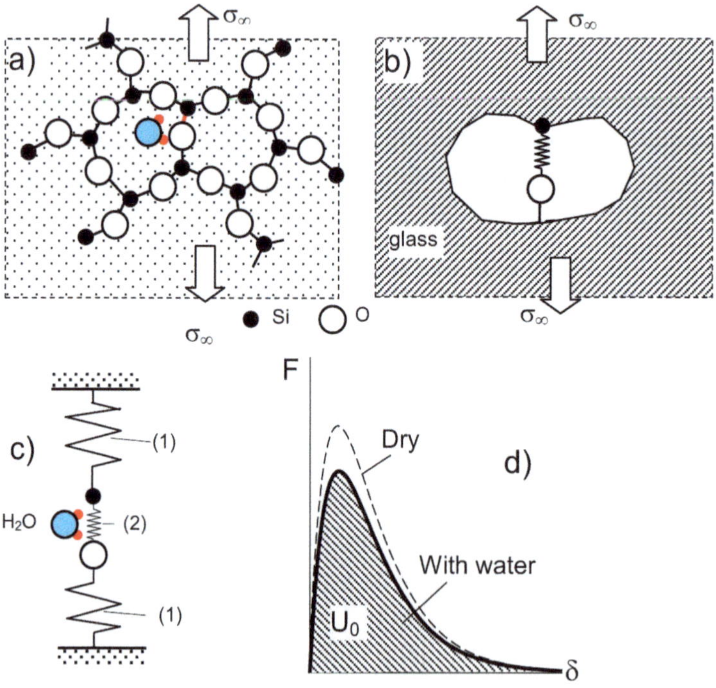

Fig. A4.1 a) Model of the loading situation of a stretched bond in the glass network (schematically), b) silica network replaced by a continuum, linear springs (1): elastic surrounding, non-linear spring (2): interaction forces between two bond partners, c) elastic response of silica network replaced by linear springs, reacting bond modelled by a nonlinear spring, d) Morse-type force-displacement curves.

The critical stress in eq.(A4.1.2) will be equal to $\sigma_c = \sigma_0$ in the absence of water where σ_0 is the ideal or theoretical strength. For σ_c see for instance [A4.6, A4.7] and for σ_0 [A4.8, A4.9]. For a small external load (Fig. A4.2a) the bond is displaced by the amount $\delta^{(1)}$, which represents the state of equilibrium. Even when thermal activation events with high energies occur, no additional state of equilibrium will be possible, and the bond will be re-established without a new thermal event. In this case the bond cannot be broken permanently.

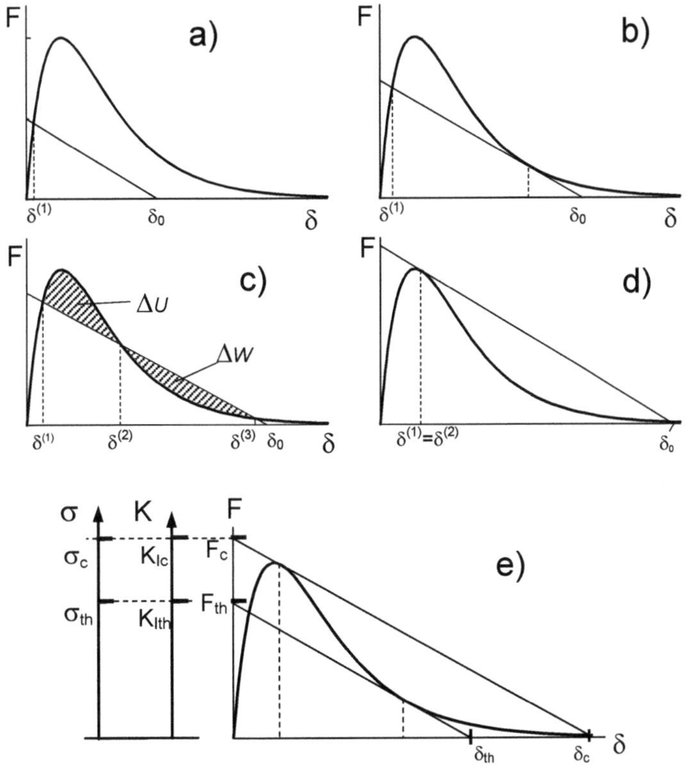

Fig. A4.2 Load–displacement behaviour of an individual bond: a)–d) increasing load, e) definition of critical and threshold forces, stress intensity factors and stresses.

At an increased external load (Fig. A4.2b) there exists a second displacement where the straight line and the Morse curve have a common tangent. At a higher load (Fig. A4.2c) we find three intersections of the two curves. If in this case a thermal activation event with energy of at least ΔU occurs, the bond will find a new position of equilibrium at $\delta^{(3)}$, and the crack can extend by a lattice distance. But this new position with its negligible bond interactions is not necessarily a permanent one. If further thermal events with energies of at least ΔW appear at the considered location, which lead to a decrease in the distance of the bond partners, the bond can be re-established.

A4.2 Bond-fracture frequency

The probability of occurrence of thermal activation events with energies high enough for bond breaking is described by the Boltzmann distribution, namely

$$\frac{\overline{N}}{N_0} = \exp(-\Delta U / kT), \qquad \text{(A4.2.1)}$$

In this expression $\overline{N} / N_0$ (N_0=number of bonds) is the fraction of events with energies greater than the energy barrier ΔU and k is Boltzmann's constant. For describing the frequency of bond-fracture events knowledge of the potential barriers ΔU and ΔW is necessary. Numerical computations are reported in [A4.2].
During a time span of Δt the number of cracked bonds ΔN is in a statistical sense

$$\Delta N / N_0 = \exp(-\Delta U / kT)[1 - \exp(-\Delta W / kT)] \qquad \text{(A4.2.2)}$$

From (A4.2.1) it can be seen that as the stress on a crack decreases, the expression in brackets [ ] becomes zero, and the share of open bonds vanishes. The related stress value is the threshold stress σ_{th}.

A decrease in the Young´s modulus is to be expected from the fact that broken bonds cannot transfer any or at least greatly reduced loads. The consequence is a decrease in the tensile stresses in the glass.

References

A4.1 Fett, T.: A fracture-mechanical theory of subcritical crack growth in ceramics, Int. J. Fract. **54**(1992), 117–130.

A4.2 Fett, T. (1992): Interpretation of subcritical crack growth in ceramic materials, DFVLR-Forschungsbericht, 82-07, 1982, (in German).

A4.3 Munz, D., Fett, T., Ceramics, Failure, Material Selection, Design. Springer, Berlin Heidelberg New York (1999).

A4.4 Fuller, E.R., Thomson, R.M.: Lattice theories of fracture, in: Fracture Mechanics of Ceramics Vol. IV, Plenum Press, (1978), 507–548, New York.

A4.5 E.R. Fuller, R.M. Thomson, Theory of chemically assisted fracture, ICM 3, Vol.2, Cambridge, Aug. 1979, 485-494.

A4.6 Kurkjian, D. Inniss, Q Zhong, C.R. Kurkjian, Chemically corroded pristine silica fibers; Blunt or sharp flaws?, J. Am. Ceram. Soc. **76**(1993), 3173-77.

A4.7 C.R. Kurkjian, P.K. Gupta, R.K. Brow, and N. Lower, "The intrinsic strength and fatigue of oxide glasses," *J. Noncrystal. Solids*, **316** 114-124 (2003).

A4.8 Silva, E. C. C. M.; Tong, L.; Yip, S.; Van Vliet, K. J. *Small* **2006**, *2*, 239–243.

A4.9 G. Brambilla, D.N. Payne, The ultimate strength of glass silica nanowires, Nano Letters, **9**(2009), 831-835.

FINAL REMARKS AND OUTLOOK

Extremely high stresses at crack tips and the water entrance in the case of humid environments cause high hydroxyl concentrations (Sections 5, 6) accompanied by large volume swelling and strong damage. In this Section, a first insight into the problematic nature may be tried for cracks and slender notches applying purely elastic material behaviour. In actual studies, the authors investigate the theme by:

a) Use of a micro-structurally motivated approach, where the crack tip regions are considered as slender notches with root radii in the order of the average radius of the SiO_2 rings,

b) Alternatively by application of well-established models borrowed from Elastic-Plastic Fracture Mechanics (EPFM).

c) In our considerations on fiber strength, we made our predictions for the high-temperature range of >450°C. In this region, the equilibrium constant is very large, so that molecular water could hardly contribute to the total amount of water. For predictions in the low-temperature region, we had assumed so far that the molecular water in the glass is located in molecular size holes in the silica glass network where the water does not contribute to the volume of the glass. Consequently, molecular water should not contribute stress effects, with the consequence of $\varepsilon_C=0$. To our knowledge, no trustworthy measurements of volume changes by molecular water $\varepsilon_C\neq0$ are actually available. Therefore, we assumed disappearing volume changes by molecular water. In terms of molar units for S and C, this relation would read in the simplest case of a mixing rule in the form of eq.(2.1.5b).

In a recent paper by Garofalini et al. [1] some first results on this problem were obtained by MD-computations. From this paper the volume changes for a water concentration of C_w=5 mol-% were taken and re-plotted in Fig. 1a. On the abscissa the hydroxyl concentrations are normalized on their maximum value of $S^{(max)}=2C_w$ and the ordinates on the value ε_S obtained for the volume strain at $S=2C_w$. An approximate description reads

$$\frac{\varepsilon_v}{\varepsilon_S} = \frac{S}{2C_w} - 0.27\exp\left[-7.4\frac{S}{2C_w}\right] \tag{1}$$

For $S/(2C_w) \geq 0.25$, the data may be represented by a straight line as introduced as the dash-dotted line. The related description is given by the first term in eq.(1), i.e. by

$$\frac{\varepsilon_v}{\varepsilon_S} \approx \frac{S}{2C_w} \quad \text{for} \quad S/(2C_w) \geq 0.25 \tag{2}$$

Finally, Fig. 1b represents the same data by using the ratio S/C as the abscissa. The intersection of the curve with the abscissa is found at S/C=0.22. According to eq.(1.1.3) this corresponds to an estimated temperature of θ=-14°C. The straight line approximation for $S/(2C_w)\geq 0.25$ indicates that the temperature should be θ>45°C for the simplified description. Summarising these results via molecular-dynamics calculations we can conclude, that an effect of molecular water is only of interest for temperatures near room temperature and can be ignored in all cases considered in this booklet. On the other hand, it should be noted that MD calculations can never replace missing experimental results. However, they allow at least qualitative statements about trends.

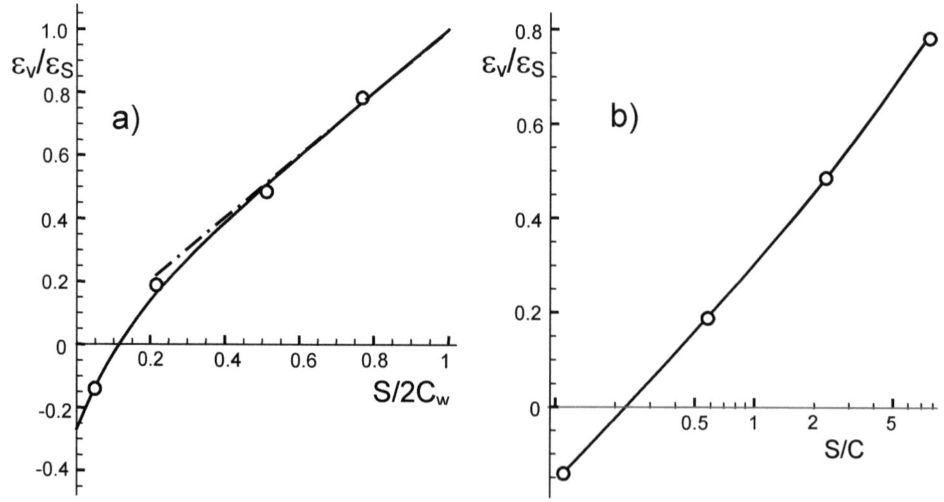

Fig. 1 a) Relative volume strain vs. relative hydroxyl concentration (curve: eq.(1)), dash-dotted line according to eq.(2), b) relative volume strain vs. equilibrium constant S/C.

d) In some experiments, the effects of swelling on samples soaked in liquid water were significantly stronger than on samples in water vapour at the same elevated temperature and saturation pressure. In Fig. 2a, the predictions on the basis of hydroxyl concentrations in liquid water, derived from measurements of total water concentration by Zouine et al. [2], are shown as the blue circles. The red squares indicate the results from the disk tests in Section 2. From measurements of surface swelling stresses from crack-terminating angles in DCDC-specimens [3, 4], the blue upright triangles were obtained. Inert strength measurements on water- and vapour-soaked bending bars, [5], yielded the black downward facing triangles. In all of the results, the values under vapour pressure are below those in water. It is somewhat astonishing, that at 90°C the two results are identical. This is the reason why the data points were slightly shifted from one another with respect to the temperature axis.

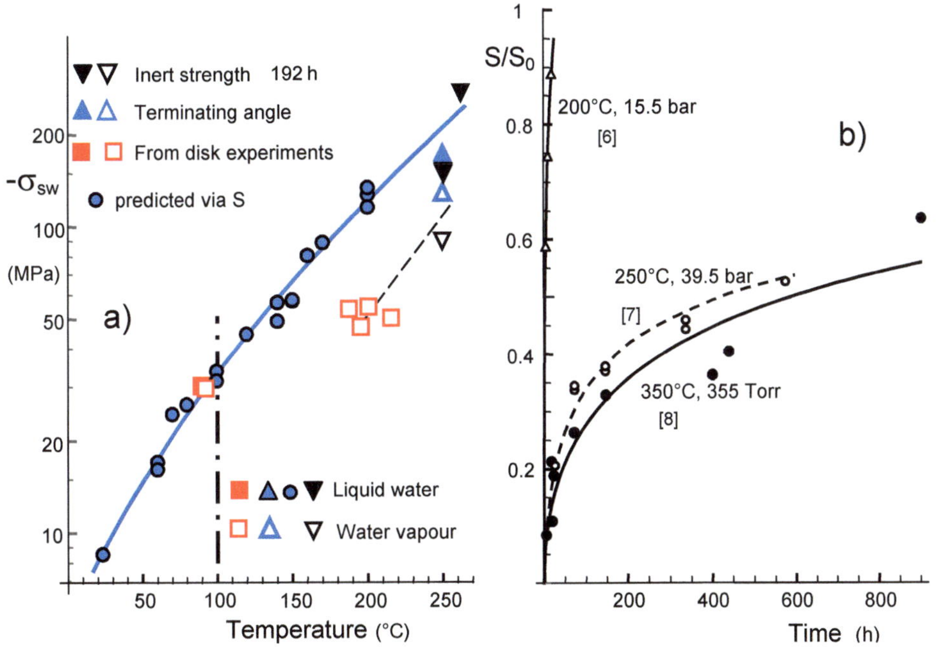

Fig. 2 a) Predicted (blue circles) and indirectly measured swelling stress results from disk curvature (red squares), terminating angles (blue triangles), and inert strengths (black downward triangles). The squares for 90°C are slightly displaced, b) Hydroxyl water concentration *S* normalized on saturation value S_0 vs. time.

Figure 2b represents again results from Section 1. The hydroxyl concentrations for measurements in water vapour are plotted in normalized form with the ordinate S normalized on the individual saturation value S_0 obtained by extrapolation to $t\rightarrow\infty$.

As a clear trend it can be seen that the normalized surface concentrations S/S_0 for the same time decrease with increasing temperature. However, it should be noted that in addition to the temperatures, different pressures and glasses with possibly different properties were used. Whereas the results by Helmich and Rauch [6] (triangles) show saturation after about 50-100 h, saturation for the results by Oehler and Tomozawa [7] at 250°C/39.5 bar and by Davis and Tomozawa [8] at 350°C/355 Torr is significantly not reached below 1000 h vapour soaking. The reason for these differences is assumed to be affected by a variation of mass-transfer conditions at the surface (see Sections 1.2 and 1.3). This topic has to be deepened in future work.

References

1 S.H. Garofalini, J. Lentz, M. Homann, Molecular mechanism of the expansion of silica glass upon exposure to moisture, J. Am. Ceram. Soc. **103** (2020), 2421-2431.

2 A. Zouine, O. Dersch, G. Walter and F. Rauch, "Diffusivity and solubility of water in silica glass in the temperature range 23-200°C," *Phys. Chem. Glass: Eur. J. Glass Sci and Tech. Pt. B*, **48** [2] (2007), 85-91.

3 P. Hettich, K. G. Schell, G. Rizzi, S. Wagner, T. Fett, Evaluation of crack-terminating angles in heat-treated silica DCDC-specimens, Scientific Working Papers **114**, 2019, ISSN: 2194-1629, Karlsruhe, KIT.

4 P. Hettich, K. G. Schell, G. Rizzi, S. Wagner, T. Fett, Estimation of swelling stresses from crack-terminating angles, Scientific Working Papers **119**, 2019, ISSN: 2194-1629, Karlsruhe, KIT.

5 P. Hettich, S. Wagner, K.G. Schell, T. Fett, Inert strength measurement on silica soaked at 250°C in liquid water and water vapour, Scientific Working Papers **122**, 2019, ISSN: 2194-1629, Karlsruhe, KIT.

6 M. Helmich and F. Rauch, On the mechanism of diffusion of water in silica glass, *Glastech. Ber.* **66** [8] (1993), 195-200.

7 Oehler, A., Tomozawa, M., Water diffusion into silica glass at a low temperature under high water vapor pressure, J. Non-Cryst. Sol. **347** (2004) 211-219.

8 Davis, K.M., Tomozawa, M., Water diffusion into silica glass: structural changes in silica glass and their effect on water solubility and diffusivity, J. Non-Cryst. Sol. **185** (1995), 203-220.

Schriftenreihe des Instituts für Angewandte Materialien

ISSN 2192-9963

Band 1 Prachai Norajitra
Divertor Development for a Future Fusion Power Plant.
ISBN 978-3-86644-738-7

Band 2 Jürgen Prokop
Entwicklung von Spritzgießsonderverfahren zur Herstellung von Mikrobauteilen durch galvanische Replikation.
ISBN 978-3-86644-755-4

Band 3 Theo Fett
New contributions to R-curves and bridging stresses – Applications of weight functions.
ISBN 978-3-86644-836-0

Band 4 Jérôme Acker
Einfluss des Alkali/Niob-Verhältnisses und der Kupferdotierung auf das Sinterverhalten, die Strukturbildung und die Mikrostruktur von bleifreier Piezokeramik $(K_{0,5}Na_{0,5})NbO_3$.
ISBN 978-3-86644-867-4

Band 5 Holger Schwaab
Nichtlineare Modellierung von Ferroelektrika unter Berücksichtigung der elektrischen Leitfähigkeit.
ISBN 978-3-86644-869-8

Band 6 Christian Dethloff
Modeling of Helium Bubble Nucleation and Growth in Neutron Irradiated RAFM Steels.
ISBN 978-3-86644-901-5

Band 7 Jens Reiser
Duktilisierung von Wolfram. Synthese, Analyse und Charakterisierung von Wolframlaminaten aus Wolframfolie.
ISBN 978-3-86644-902-2

Band 8 Andreas Sedlmayr
Experimental Investigations of Deformation Pathways in Nanowires.
ISBN 978-3-86644-905-3

Die Bände sind unter www.ksp.kit.edu als PDF frei verfügbar oder als Druckausgabe bestellbar.

Band 9 Matthias Friedrich Funk
Microstructural stability of nanostructured fcc metals during cyclic deformation and fatigue.
ISBN 978-3-86644-918-3

Band 10 Maximilian Schwenk
Entwicklung und Validierung eines numerischen Simulationsmodells zur Beschreibung der induktiven Ein- und Zweifrequenzrandschichthärtung am Beispiel von vergütetem 42CrMo4.
ISBN 978-3-86644-929-9

Band 11 Matthias Merzkirch
Verformungs- und Schädigungsverhalten der verbundstranggepressten, federstahldrahtverstärkten Aluminiumlegierung EN AW-6082.
ISBN 978-3-86644-933-6

Band 12 Thilo Hammers
Wärmebehandlung und Recken von verbundstranggepressten Luftfahrtprofilen.
ISBN 978-3-86644-947-3

Band 13 Jochen Lohmiller
Investigation of deformation mechanisms in nanocrystalline metals and alloys by in situ synchrotron X-ray diffraction.
ISBN 978-3-86644-962-6

Band 14 Simone Schreijäg
Microstructure and Mechanical Behavior of Deep Drawing DC04 Steel at Different Length Scales.
ISBN 978-3-86644-967-1

Band 15 Zhiming Chen
Modelling the plastic deformation of iron.
ISBN 978-3-86644-968-8

Band 16 Abdullah Fatih Çetinel
Oberflächendefektausheilung und Festigkeitssteigerung von niederdruckspritzgegossenen Mikrobiegebalken aus Zirkoniumdioxid.
ISBN 978-3-86644-976-3

Band 17 Thomas Weber
Entwicklung und Optimierung von gradierten Wolfram/ EUROFER97-Verbindungen für Divertorkomponenten.
ISBN 978-3-86644-993-0

Band 18 Melanie Senn
Optimale Prozessführung mit merkmalsbasierter Zustandsverfolgung.
ISBN 978-3-7315-0004-9

Band 19 Christian Mennerich
Phase-field modeling of multi-domain evolution in ferromagnetic shape memory alloys and of polycrystalline thin film growth.
ISBN 978-3-7315-0009-4

Band 20 Spyridon Korres
On-Line Topographic Measurements of Lubricated Metallic Sliding Surfaces.
ISBN 978-3-7315-0017-9

Band 21 Abhik Narayan Choudhury
Quantitative phase-field model for phase transformations in multi-component alloys.
ISBN 978-3-7315-0020-9

Band 22 Oliver Ulrich
Isothermes und thermisch-mechanisches Ermüdungsverhalten von Verbundwerkstoffen mit Durchdringungsgefüge (Preform-MMCs).
ISBN 978-3-7315-0024-7

Band 23 Sofie Burger
High Cycle Fatigue of Al and Cu Thin Films by a Novel High-Throughput Method.
ISBN 978-3-7315-0025-4

Band 24 Michael Teutsch
Entwicklung von elektrochemisch abgeschiedenem LIGA-Ni-Al für Hochtemperatur-MEMS-Anwendungen.
ISBN 978-3-7315-0026-1

Band 25 Wolfgang Rheinheimer
Zur Grenzflächenanisotropie von $SrTiO_3$.
ISBN 978-3-7315-0027-8

Band 26 Ying Chen
Deformation Behavior of Thin Metallic Wires under Tensile and Torsional Loadings.
ISBN 978-3-7315-0049-0

Die Bände sind unter www.ksp.kit.edu als PDF frei verfügbar oder als Druckausgabe bestellbar.

Band 27 Sascha Haller
Gestaltfindung: Untersuchungen zur Kraftkegelmethode.
ISBN 978-3-7315-0050-6

Band 28 Nicht erschienen

Band 29 Gunnar Picht
Einfluss der Korngröße auf ferroelektrische Eigenschaften dotierter $Pb(Zr_{1-x}Ti_x)O_3$ Materialien.
ISBN 978-3-7315-0106-0

Band 30 Esther Held
Eigenspannungsanalyse an Schichtverbunden mittels inkrementeller Bohrlochmethode.
ISBN 978-3-7315-0127-5

Band 31 Pei He
On the structure-property correlation and the evolution of Nanofeatures in 12-13.5% Cr oxide dispersion strengthened ferritic steels.
ISBN 978-3-7315-0141-1

Band 32 Jan Hoffmann
Ferritische ODS-Stähle – Herstellung, Umformung und Strukturanalyse.
ISBN 978-3-7315-0157-2

Band 33 Wiebke Sittel
Entwicklung und Optimierung des Diffusionsschweißens von ODS Legierungen.
ISBN 978-3-7315-0182-4

Band 34 Osama Khalil
Isothermes Kurzzeitermüdungsverhalten der hoch-warmfesten Aluminium-Knetlegierung 2618A (AlCu2Mg1,5Ni).
ISBN 978-3-7315-0208-1

Band 35 Nicht erschienen

Band 36 Christoph Hage
Grundlegende Aspekte des 2K-Metallpulverspritzgießens.
ISBN 978-3-7315-0217-3

Band 37 Bartłomiej Albiński
Instrumentierte Eindringprüfung bei Hochtemperatur für die Charakterisierung bestrahlter Materialien.
ISBN 978-3-7315-0221-0

Band 38 Tim Feser
Untersuchungen zum Einlaufverhalten binärer alpha-Messinglegierungen unter Ölschmierung in Abhängigkeit des Zinkgehaltes.
ISBN 978-3-7315-0224-1

Band 39 Jörg Ettrich
Fluid Flow and Heat Transfer in Cellular Solids.
ISBN 978-3-7315-0241-8

Band 40 Melanie Syha
Microstructure evolution in strontium titanate Investigated by means of grain growth simulations and x-ray diffraction contrast tomography experiments.
ISBN 978-3-7315-0242-5

Band 41 Thomas Haas
Mechanische Zuverlässigkeit von gedruckten und gasförmig abgeschiedenen Schichten auf flexiblem Substrat.
ISBN 978-3-7315-0250-0

Band 42 Aron Kneer
Numerische Untersuchung des Wärmeübertragungsverhaltens in unterschiedlichen porösen Medien.
ISBN 978-3-7315-0252-4

Band 43 Manuel Feuchter
Investigations on Joule heating applications by multiphysical continuum simulations in nanoscale systems.
ISBN 978-3-7315-0261-6

Band 44 Alexander Vondrous
Grain growth behavior and efficient large scale simulations of recrystallization with the phase-field method.
ISBN 978-3-7315-0280-7

Band 45 Tobias Kennerknecht
Fatigue of Micro Molded Materials – Aluminum Bronze and Yttria Stabilized Zirconia.
ISBN 978-3-7315-0293-7

Band 46 Christopher Scherr
Elektrochemisches Verhalten von Lithium-Schwefel-Zellen mit unterschiedlicher Kathodenstruktur.
ISBN 978-3-7315-0296-8

Band 47 Konstantin Frölich
Der Decal-Prozess zur Herstellung katalysatorbeschichteter Membranen für PEM-Brennstoffzellen.
ISBN 978-3-7315-0334-7

Band 48 Benedikt Haspel
Werkstoffanalytische Betrachtung der Eigenschaften von mittels neuartiger RTM-Fertigungsprozesse hergestellten glasfaserverstärkten Polymerverbunden.
ISBN 978-3-7315-0337-8

Band 49 Marco Berghoff
Skalenübergreifende Modellierung und Optimierung vom atomistischen kristallinen Phasenfeldmodell bis zur mesoskopischen Phasenfeldmethode.
ISBN 978-3-7315-0416-0

Band 50 Michael Selzer
Mechanische und Strömungsmechanische Topologieoptimierung mit der Phasenfeldmethode.
ISBN 978-3-7315-0431-3

Band 51 Michael Mahler
Entwicklung einer Auswertemethode für bruchmechanische Versuche an kleinen Proben auf der Basis eines Kohäsivzonenmodells.
ISBN 978-3-7315-0441-2

Band 52 Christoph Bohnert
Numerische Untersuchung des Verformungs- und Bruchverhaltens von einkristallinem Wolfram auf mikroskopischer Ebene.
ISBN 978-3-7315-0444-3

Band 53 Stefan Guth
Schädigung und Lebensdauer von Nickelbasislegierungen unter thermisch-mechanischer Ermüdungsbeanspruchung bei verschiedenen Phasenlagen.
ISBN 978-3-7315-0445-0

Band 54 Markus Klinsmann
The Effects of Internal Stress and Lithium Transport on Fracture in Storage Materials in Lithium-Ion Batteries.
ISBN 978-3-7315-0455-9

Band 55 Thomas Straub
Experimental Investigation of Crack Initiation in Face-Centered Cubic Materials in the High and Very High Cycle Fatigue Regime.
ISBN 978-3-7315-0471-9

Band 56 Maren Lepple
Kupfer- und Eisenoxide als Konversions-Elektrodenmaterialien für Lithium-Ionen-Batterien: Thermodynamische und Elektrochemische Untersuchungen.
ISBN 978-3-7315-0482-5

Band 57 Stefan Andreas Slaby
Charakterisierung und Bewertung der Zug- und Ermüdungseigenschaften von Mikrobauteilen aus 17-4PH Edelstahl. Ein Vergleich von mikropulverspritzgegossenem und konventionell hergestelltem Material.
ISBN 978-3-7315-0484-9

Band 58 Kumar Ankit
Phase-field modeling of microstructural pattern formation in alloys and geological veins.
ISBN 978-3-7315-0491-7

Band 59 Kuo Zhang
Characterization and Modeling of the Ratcheting Behavior of the Ferritic-Martensitic Steel P91.
ISBN 978-3-7315-0503-7

Band 60 Nicht erschienen

Band 61 Fabian Lemke
Untersuchung des Sinterverhaltens von $SrTiO_3$ unter Berücksichtigung der Defektchemie.
ISBN 978-3-7315-0510-5

Band 62 Johannes Kümmel
Detaillierte Analyse der Aufbauschneidenbildung bei der Trockenzerspanung von Stahl C45E mit Berücksichtigung des Werkzeugverschleißes.
ISBN 978-3-7315-0518-1

Band 63 László Hagymási
Modellierung der Stoffübertragung beim Niederdruckcarbonitrieren mit Ammoniak und Acetylen.
ISBN 978-3-7315-0568-6

Die Bände sind unter www.ksp.kit.edu als PDF frei verfügbar oder als Druckausgabe bestellbar.

Band 64 Reza Eslami
A novel micro-mechanical model for prediction of multiaxial high cycle fatigue at small scales.
ISBN 978-3-7315-0583-9

Band 65 Sebastian Schulz
Phase-field simulations of multi-component solidification and coarsening based on thermodynamic datasets.
ISBN 978-3-7315-0618-8

Band 66 Markus Stricker
Die Übertragung von mikrostrukturellen Eigenschaften aus der diskreten Versetzungsdynamik in Kontinuumsbeschreibungen.
ISBN 978-3-7315-0658-4

Band 67 Luis Straßberger
Untersuchung und Modellierung des viskoplastischen Verformungsverhaltens oxidpartikelverstärkter Stähle.
ISBN 978-3-7315-0674-4

Band 68 Mark Wobrock
Microplasticity of idealized single crystalline Ag cantilevers characterized with methods of high resolution.
ISBN 978-3-7315-0682-9

Band 69 Amritesh Kumar
Micromechanical study on the deformation behaviour of directionally solidified NiAl-Cr eutectic composites.
ISBN 978-3-7315-0694-2

Band 70 Johannes Hötzer
Massiv-parallele und großskalige Phasenfeldsimulationen zur Untersuchung der Mikrostrukturentwicklung.
ISBN 978-3-7315-0693-5

Band 71 Thomas Hupfer
Herstellung von LATP für den Einsatz als Festkörperelektrolyt und dessen Eigenschaften.
ISBN 978-3-7315-0702-4

Band 72 Florentin Pottmeyer
Schädigungsverhalten von in CFK-Laminaten eingebetteten Inserts unter bauteilnahen Beanspruchungen.
ISBN 978-3-7315-0719-2

Band 73 Andres Höweling
Untersuchung der Hochvoltstabilität und Tiefentladung von dotierten $LiNi_{0,5}Mn_{1,5}O_4$-Hochvoltspinellen.
ISBN 978-3-7315-0728-4

Band 74 Tabea Gisela Schwark
Deformation and Fracture Properties of the Soft Magnetic Composite Somaloy 700 3P on Different Length Scales.
ISBN 978-3-7315-0759-8

Band 75 Klaudia Lichtenberg
Metallmatrixverbunde mit Verstärkungselementen aus metallischem Glas $Ni_{60}Nb_{20}Ta_{20}$ – Herstellung und Charakterisierung.
ISBN 978-3-7315-0782-6

Band 76 Claudio Findeisen
Charakterisierung und Modellierung von instabilen Metamaterialien.
ISBN 978-3-7315-0869-4

Band 77 Nilesha Mishra
Influence of strain on the functionality of ink-jet printed thin films and devices on flexible substrates.
ISBN 978-3-7315-0853-3

Band 78 Simon Werner Bonk
Plastische Verformungsmechanismen in hochgradig kaltgewalzten, ultrafeinkörnigen Wolframblechen.
ISBN 978-3-7315-0878-6

Band 79 Tim Gräning
Herstellung, Charakterisierung und Optimierung von austenitischen ODS Stählen.
ISBN 978-3-7315-0732-1

Band 80 Peter Rupp
Herstellung, Prüfung und Modellierung neuartiger hybrider Aluminiumschaum-CFK-Sandwichverbunde.
ISBN 978-3-7315-0880-9

Band 81 Benjamin Sebastian Ehreiser
Einfluss mechanischer Lasten auf die Herstellung von Stahl-Glaskeramik-Verbunden.
ISBN 978-3-7315-0954-7

Die Bände sind unter www.ksp.kit.edu als PDF frei verfügbar oder als Druckausgabe bestellbar.

Band 82 Hans Giel
Weiterentwicklung experimenteller Methoden zur Ermittlung thermodynamischer Werkstoffdaten von Lithium-Ionen-Batterien.
ISBN 978-3-7315-0981-3

Band 83 Anna Trauth
Characterisation and Modelling of Continuous-Discontinuous Sheet Moulding Compound Composites for Structural Applications.
ISBN 978-3-7315-0950-9

Band 84 Jonas Johannes Hüther
The Impact of Recycling on the Fibre and the Composite Properties of Carbon Fibre Reinforced Plastics.
ISBN 978-3-7315-0983-7

Band 85 Nicolas A. Mayer
Thermodynamik von Kobaltoxid Anodenmaterialien für Lithium-Ionen-Batterien und ihr elektrochemisches Verhalten.
ISBN 978-3-7315-0996-7

Band 86 Ulrich Führer
Untersuchung und Modellierung des Haltezeiteinflusses auf die zyklische Entfestigung ferritisch-martensitischer Stähle.
ISBN 978-3-7315-0837-3

Band 87 Ebru Cihan
Structure evolution in tribological interfaces studied by multilayer model alloys.
ISBN 978-3-7315-0999-8

Band 88 Markus Sudmanns
Entwicklung einer Kontinuumsbeschreibung für die Versetzungsmobilität in Versetzungsnetzwerken.
ISBN 978-3-7315-1001-7

Band 89 Tao Zhang
Phase-field Modeling of Phase Changes and Mechanical Stresses in Electrode Particles of Secondary Batteries.
ISBN 978-3-7315-1002-4

Band 90 Markus Ganser
On the Electro-Chemo-Mechanical Coupling in Solid State Batteries and its Impact on Morphological Interface Stability.
ISBN 978-3-7315-1047-5

Band 91 Michael Kellner
Modellierung mehrkomponentiger Materialsysteme für die Phasenfeldmethode und Analyse der simulierten Mikrostrukturen.
ISBN 978-3-7315-1044-4

Band 92 Felix Schröckert
Herstellung dünner Folien aus Lithium-Lanthan-Titanat zur Anwendung als Festkörperelektrolyt.
ISBN 978-3-7315-1008-6

Band 93 Ephraim Schoof
Chemomechanische Modellierung der Wärmebehandlung von Stählen mit der Phasenfeldmethode.
ISBN 978-3-7315-1050-5

Band 94 Alexander Valentin Brabänder
Registrierende Härtemessung an neutronenbestrahlten Materialien bei hohen Temperaturen.
ISBN 978-3-7315-1097-0

Band 95 Denny Schmidt
Einfluss der Kompaktierung auf die Elektrodenmikrostruktur und elektrochemische Performance bei Lithium-Ionen-Zellen.
ISBN 978-3-7315-1098-7

Band 96 Svenja Dittrich
Entwicklung von Siebdruckpasten zur Herstellung von Glaslotfügungen für die Festoxidbrennstoffzelle.
ISBN 978-3-7315-1085-7

Band 97 Michael Dippon
Bestimmung der Betriebsgrenzen für das Schnellladen von Lithium-Ionen Batterien.
ISBN 978-3-7315-1123-6

Band 98 Patricia Haremski
Diffusionseigenschaften von Nickel in einer Festoxid-Brennstoffzelle.
ISBN 978-3-7315-1124-3

Die Bände sind unter www.ksp.kit.edu als PDF frei verfügbar oder als Druckausgabe bestellbar.

Band 99 Florian Wankmüller
Mehrskalige Charakterisierung der Hochtemperatur-Brennstoffzelle (SOFC).
ISBN 978-3-7315-1142-7

Band 100 Niklas Russner
Modellgestützte Analyse des Stackbetriebs von Festoxidzellen.
ISBN 978-3-7315-1144-1

Band 101 Theo Fett, Karl Günter Schell, Ethel C. Bucharsky, Gabriele Rizzi, Pascal Hettich, Susanne Wagner, Michael J. Hoffmann
Consequences of hydroxyl generation by the silica/water reaction - Part I: Diffusion and Swelling.
ISBN 978-3-7315-1148-9

Die Bände sind unter www.ksp.kit.edu als PDF frei verfügbar oder als Druckausgabe bestellbar.